12 Birds to Save Your Life

12 Birds to Save Your Life

Nature's Lessons in Happiness

CHARLIE CORBETT

MICHAEL JOSEPH

MICHAEL JOSEPH

UK | USA | Canada | Ireland | Australia
India | New Zealand | South Africa

Michael Joseph is part of the Penguin Random House group of companies
whose addresses can be found at global.penguinrandomhouse.com.

First published 2021
001

Copyright © Charlie Corbett, 2021

The moral right of the author has been asserted

Set in 13.5/16pt Garamond MT Std
Typeset by Jouve (UK), Milton Keynes
Printed and bound in Great Britain by Clays Ltd, Elcograf S.p.A.

The authorized representative in the EEA is Penguin Random House Ireland,
Morrison Chambers, 32 Nassau Street, Dublin D02 YH68

A CIP catalogue record for this book is available from the British Library

HARDBACK ISBN: 978–0–241–50333–1

For Mum

because:

'*Oh, I don't know, you know, don't you know?*'

. . . as Plum once said.

Lend me not to another and I will be a quiet companion in all your wanderings. Wherever thou goest there go I, through the eagle's air and over the wide seas; through heat and cold, calm and tempest, and the changing years. When thou layest thyself down upon thy bed when the weary day is over read of me a little and thy dreams shall be sweet.

Isaak Walton, The Compleat Angler

Speak, roofless Nature, your instinctive words;
And let me learn your secret from the sky

Siegfried Sassoon

Introduction

This book is about coping with being human. It's about how unexpected events smashed and bashed their way into my life without permission. And how a reinvigorated love of the natural world – in particular its birds – helped me to cope with the death of my mother and anchor myself at a time of great uncertainty and anxiety. There is nothing new about death and change, anxiety and melancholy. They have been with us since our ancestors first stepped out of the primordial swamp and worried about what they were going to eat for dinner. And yet for some reason, as a species, we have become worse at dealing with these traits than ever before.

We have become dislocated from our natural environment. And it is this dislocation, I believe, that is a major reason why so many of us now find it hard to cope with what the modern world throws at us. We've lost the perspective that nature provides.

Not so very long ago, knowledge of something as straightforward as the birds around us was innate. It was a part of being human, like having legs and arms. Knowing about birds, and nature, was not just the hobby of a few thousand eccentric individuals who baffle us by racing around the country looking for rare species they mark up in a special book. Bird lovers are put on a par, these days, with trainspotters and other obsessive hobbyists: harmless, but slightly odd.

And until recently I too knew next to nothing about the wildlife around me. After many years of living in cities at home and abroad, I returned to the small farm where I grew up and felt like an alien. I walked along the lanes and byways of my childhood and found I was surrounded by strangers. Random birds in unnamed trees singing songs I couldn't understand. I realized that my store of knowledge, garnered from a childhood in rural England, had largely disappeared. I could barely tell the difference between a sparrow and a starling. And as for the dawn chorus, it baffled me. I felt ashamed, and set out to put things right.

In this journey of rediscovery I came to realize that if you take note of the nature around you, try to understand it, watch it, and grow to love it, then not only will this make you a happier, more content human being, but it will help nature too. It will move it from the realm of the abstract into the world of the tangible.

The benefits I derived from reconnecting with the birds around me during the time of my mother's illness, death and the painful aftermath were incalculable. I garnered much-needed mental ballast. And I want you, through reading this book, to feel the same unadulterated joy I feel when I hear a blackbird's song on a fresh spring morning, a song thrush singing at the end of the longest night, or when I watch the house martins arrive from Africa in April. It is life-saving.

Some of the twelve birds in this book will be more

familiar to you than others. I've tried to select birds that not only mean a great deal to me – and played a part in my mental rehabilitation – but which are also birds that live all around you right now – or at least not far away – that you can very easily go and see for yourself. I have also added simple descriptions of each bird, and the places you are most likely to come across it, at the foot of each chapter. And at the end of the book, to add context, I've written a short gazetteer that details which birds you are likely to see at different times of the year, and in different places.

Reading this book will not make you an expert in birds, or able to identify unusual flora and fauna. My aim is to help you rebuild your relationship with the natural world around you. To become grounded. And to put your life and troubles into perspective by learning to understand, love and begin to move with the rhythms of the natural world outside your door.

Some days, when the shadow of the black dog grows particularly large, because it still does, I will wander up the side of a nearby hill and simply lie down in the grass. If I'm lucky, the skylarks will be singing. Or maybe a mob of yellowhammers will bustle past, or even a charm of goldfinches (the mere fact that the collective noun for goldfinches is a 'charm' warms my heart). Taking time to pay attention to nature, to be among nature, always has the effect of putting my problems into perspective. And it teaches me lessons about myself, and life, each and every day.

I hope this book acts as a reminder to people that

there is an entire other dimension to their lives that's just waiting to be acknowledged. Once you do, and begin to try to understand it, then even the very smallest events can bring such joy.

Prologue

In my beginning is my end

I am driving across the roof of Scotland – west to east over the Cairngorms – on a clear, undulating road. Above me is a cloudless sky. Next to me is my girlfriend, Mary. On the radio is the Who, 'Baba O'Riley' playing at full volume. Even though it is sunny and midsummer, the temperature gauge on my car reads just ten degrees. We're thousands of feet up in the sky. Laughing. The windows are wide open and the air is cool. We're on our way to a party. On our way back from another party. The future is unwritten. I am happy. We are happy. I'm going to marry this girl in December. And life tastes sweet.

1. Skylark

The lark's on the wing;
The snail's on the thorn:
God's in his Heaven—
All's right with the world!
        Robert Browning

(Like to the lark at the break of day arising)
From sullen earth sings hymns at Heaven's gate.
        William Shakespeare

There is no better word to describe the feeling you have when you hear and see a skylark than 'exaltation'. And I'm not alone in thinking that. 'Exaltation' is the collective noun for a group of skylarks. And it's spot on. I've absolutely no idea who it was that came up with all these collective nouns back in the day (a clattering of jackdaws, a murmuration of starlings, a charm of goldfinches – to name but a few of the more glorious ones). But whoever it was, was bloody good at their job.

The song of the skylark has genuine power. It can, and will, lift your spirits no matter what is happening in your wider life. Quite apart from the fact that in order to hear a skylark you will need to be in a wide-open and wind-blown spot, devoid of people, pollution and concrete – a good source of happiness on its own. Then the rolling, tinkling, soaring song of the lark will transport you to an even better place. It is very hard to describe the sense of inner tranquillity I feel when I hear those notes. When I look up and see a skylark soaring above, hovering absolutely still above my head – wings flapping furiously – as it pours out its little heart.

If it were a drink, the impact on your soul of hearing a skylark's song would be like that first ice-cold gin and tonic after a hot and stressful day in the office. No. Scratch that. The G & T doesn't do it justice. More like that precise moment you dip your body into a deliciously warm bath after a bone-achingly difficult day on your feet. That's getting close, I think. But we're not quite there yet.

Perhaps a better analogy altogether would be to compare lark song to the Buddhist's butterfly. You can spend your life searching for happiness and very likely never find it. Stop looking, your friendly local Buddhist will tell you. Happiness is like a butterfly. You can chase it all day and never capture it. It's only when you give up and sit down to rest and reflect that the butterfly lands gently on your shoulder, without you even noticing. Well, lark song is the musical equivalent. It's always been in the background; you just need to stop, take a breath, and listen out for it.

The skylark played a central role in my own birdsong epiphany. It lifted me in a dark hour, in a way nothing man-made could ever have achieved. And it reinforced in me an awestruck wonder at the power of nature that has never left me.

It had been one of those days. Days that crop up every so often in a lifetime that start well – all sunshine, optimism and vim – but end in disarray. My family had heard news we thought we'd never hear. Ever. Perhaps we'd been in some kind of collective denial for too long. That strange human trait to wilfully ignore all the facts, right up to the moment when the facts jump all over the breakfast table, pee in your cornflakes and then force-feed themselves to you with a rusty spoon. I think the word for it is 'hope'. And it brings to mind that wonder-ful quote from John Cleese in the film *Clockwise*: 'It's not the despair . . . I can stand the despair. It's the *hope*!' (It's much funnier when he says it.)

Anyway, it was on that day – a dog day in August – that we, as a family, got our bad news. When our hope was speared through the heart. My mother, who'd been diag-nosed with a brain tumour a month before, was issued with a dreaded time limit. For the first time in my life words like 'malignant', 'terminal' and 'palliative care' were used in direct reference to someone I loved. They bounced off my skull as the consultant uttered them. Nothing sank in. It felt otherworldly and unreal. Nothing to do with me or my mum. I remember looking across at my family. There were five of us in that little cell of a room with its tin-rattled blinds and institutional grey car-pets. Mum and Dad seated together, conjoined, my elder sister, Katie, clutching Mum's hand like she would never let go, my brother, Richard, his hand resting softly on Mum's shoulder. And then there was me, the youngest, at the back, trying to piece it all together. Each of us, in turn, looked at the other. More than anything else, our expressions revealed utter confusion.

Up to that point, you see, we'd genuinely believed that what Mum was going through was merely a blip in an otherwise long and fulfilling life. That we'd clear this fence, our mother would recover, we'd all learn a few valuable lessons, and life would go on just as it had before. No hard feelings. In the future we'd laugh about this strange time, and Dad would moan about how much it had cost to park in an English hospital car park.

It had all started one warm July morning a month before. My wife, Mary, and I were at the kitchen table, happily planning our weekend, when the phone rang. It was Dad. 'I'm a little worried about your mother,' he said. 'It's likely nothing at all, but she's had a funny turn.'

'What? A funny turn? What does that mean, Dad?'

'Well, just that. She was peeling boiled eggs for a picnic and she had this funny turn. We're taking her to the doctor, just to be safe.'

A funny turn. I put the phone down and got on with my life. Mental note to self: *Give Mum a call later to see how she is.* And that was that. But, of course, it wasn't that. It turned out that Mum had been having headaches for weeks on end without telling anyone, and the doctor was worried enough to arrange further scans. But still, it was probably nothing, we all agreed, probably nothing. Our mother was the fittest and youngest sixty-six-year-old we all knew. She oozed an energetic cheerfulness out of every pore. People like Mum don't get ill, we reasoned. It really is as simple as that. She had enough in the tank to get her well beyond a hundred years, and a bit further than that, too. And it helped that she too was firmly of this view. 'I'm absolutely fine, darlings. It really is probably nothing. Just a funny turn.'

And so the days turned into weeks. And that sunny July day faded into memory. We existed in a limbo, where Mum went through test after test, consultation piled upon consultation, as we tried to get to the bottom of her condition. We'd known it was some kind of tumour in her head, but it appeared that there were lots of different types of these horrible things. Some were eminently operable, from which you could recover completely, while others had more sinister connotations. The word we kept clinging on to was 'benign'. Not long ago Richard brought up those days: the days of confused bewilderment before the Grim Diagnosis – our Phoney War.

'Mum mentioned to me how happy she was in those early days,' he said.

I couldn't quite comprehend this comment.

'How could anyone possibly be happy at a time of such dreadful life-or-death uncertainty?'

'But don't you remember all the lunches and dinners we had together at the time? You see, for the first time in years, we were all together again as a family,' he said.

I realize now that he was right. Those days and weeks reinforced so powerfully in our minds the integral part Mum played in the creation and survival of our little family unit. And you could read in her eyes the trade-off she was making between the intense joy she felt at seeing us all together again, to be surrounded by those she loved more than anyone or anything in this world, and the trembling realization that time could very well be running out for her.

Mum was the glue that held our family together, of course. But I never truly appreciated that until it began to become unstuck. I took it for granted. We all did. And so, in those uncertain summer days – as we balanced hope with stuttering expectation, we embarked upon a kind of whirlwind family tour of the surrounding villages and towns. We trundled about the countryside as one, the five of us, in a little charabanc of cheerful trepidation: visiting doctors, dropping in at clinics, making appointments with consultants – Mum gently cradled in our hands and our hearts – and then, when the appointments were done, heading off to some local pub or hotel to have lunch. It was like a second childhood for us, the trappings of adulthood briefly cast off – our own families elsewhere. It was 1985 all over again. My father would invariably complain about the pub's menu. 'Don't you do *normal* food? All I want is a steak!' Mum rolling her eyes and apologizing to the harried staff. My sister, Katie, immovable at Mum's side – two peas in a pod – and my brother's dog, Stella, buggering off at speed at any given opportunity.

One of the very last lunches we had with Mum, in a beautiful, impossibly English picture-postcard of a pub, the George Inn, was punctuated by the loss of this serially disobedient Labrador. I shall never forget the tears of laughter in Mum's eyes as my brother tore about the village shouting after his dog, pleading with her to come back. 'Stella, Stella!! *Please* come back, Stella. Where are you, Stella? Please!!'

We searched and searched for her, and it was only as we returned to the car, hoarse, sweating and defeated, that she appeared, nonchalantly trotting down the lane without a care in the world.

'STELLA, YOU BITCH!!' my brother cried in tearful exasperation.

Stella just wagged her tail and jumped into his arms. My God, did we laugh. There was an essential humour at the heart of my family's life. No matter how grim life got, or how much we argued, or what bizarre things would go wrong, we were never more than a few heartbeats away from gales of laughter.

And even after receiving the news that Mum had a maximum of two years to live, if she was lucky, we never lost that innate ability to laugh together at our misfortunes. Even while we cried inside at the same time.

I returned home the afternoon of that Grim Diagnosis not knowing how to feel at all. Like someone had cryogenically frozen my emotions. We'd all gone our separate ways and I was alone with my thoughts. Dark thoughts. I just couldn't sit still – and restlessly paced my little kitchen, back and forth. It was like my mind had gone numb, but no one had told the rest of my body. And so I resolved to take my ill-functioning brain and head out into the empty countryside around me. What other choice did I have? I was lucky enough to be living at that time in a beautiful sequestered nook of a village embraced by the hills of the Pewsey Vale in Wiltshire, which was littered with large, empty expanses. Good walking country. I threw on a coat and wellington boots and set out into the mid-afternoon drizzle.

After some hours walking and doing my best not to think, I found myself – somewhat damp – lying on the side of a lonely hill staring up at a leaden sky, while the gentle warm August drizzle seeped into my bones.

No one trains you for receiving bad news. In the same way that no one trains – or can possibly prepare you – for having small children. Sure, they can tell you about it endlessly before you have them (and they do) but you can never actually step into those shoes until you're holding a screaming, convulsing child in the middle of a supermarket aisle in front of bemused onlookers, while your other child scampers happily out of the door with an ill-gotten bag of Maltesers and disappears towards a busy road. You never truly comprehend the word 'help-lessness' until you've been through that experience. And it's the same with bad news. There is a helplessness to the whole affair.

And, boy, did I need some training. I came from a family where you never, ever, on any account, talked about your feelings. Any show of genuine emotion, apart from laughter, of course, was met with deep embarrassment and awkwardness. Conversation swiftly moved on. And while we were a close family in many ways, we were not the sort of people who hugged. We did not discuss our 'issues' either. And we certainly didn't bore other people with our problems. This would have been regarded as self-indulgent and, quite frankly, selfish. (The fact I just put the word

'issues' in inverted commas is a strong clue as to my familial attitude towards being open about feelings.)

As children, whenever any of us asked an awkward
question, my father would blanch a bit and say, 'Ask your
mother,' before making the sharpest of exits. Similarly,
when confronted with said awkward question, my mother
too would blanch, and then respond with the same line
every time: 'Trunky want a bun?' To this day I remain
mystified by the meaning of that phrase (something to
do with hungry elephants in the zoo, I think) but it wasn't
half effective at killing awkward lines of questioning.

And so when it came to an earthquake of this magni-
tude in our previously sedate lives, we just didn't know
what to do. How to react. What to say. Who to speak to.
And that was why I found myself lying on a hill, in the
rain, numb in every way. My brain was frozen. I felt alone
in the world. Alone on this muddy old hill. And no one
could help me. Not really. Not now.

And then I heard that skylark.

The soaring, rolling, cascading music rippling through
the air above me was like an injection of neat hope to
the soul. It was a kind of airborne ecstasy being played
out above my head. In full stereo. I thought to myself,
It's August. Skylarks aren't supposed to sing in August. And yet
there it was. As plain as day. Hovering above my head,
doing what skylarks do best.

For a few joyous minutes the lark and his song transported me away from that grim day. Away from the drizzle
and the dark thoughts. It transported me to childhood
summers. Harvest times on my parents' farm. Haymaking.

The skylarks sang then, too. Glorious memories of safer, more secure times. What I am trying to say, I suppose, is that it released me.

Five minutes later I was still wet to the bone, and my dear old mum was still ill. But I had found a haven from that day's storm. Some emotional ballast with which to put on a braver face. I had been grounded by the beauty of nature. By the unexpected interruption of my dark inner monologue with a shaft of unadulterated sunshine. Hope reared its head again. And it was the song of the lark that did that.

I had been doing my best to reacquaint myself with the birds and other wildlife around me after many years of city life. But while I had done a reasonable job of learning the names and getting to know the sights and the sounds of the birds in a kind of half-hearted, box-ticking way, I had never before truly experienced the intense power nature could bestow. It was just so unexpected, that lark. In the past my appreciation of the natural world had felt very forced. I could see the beauty, but suddenly, that day, I *felt* it too. Like the Buddhist's butterfly, it had just sort of landed on my shoulder.

In the same way that I had neglected nature, as the skylark sang above me, I reflected that I had probably neglected Mum too as I had made my way in life. It's only natural, of course, to fledge the nest and roar about the world trying to carve a life for yourself. But one of the casualties of growing up and leaving home, alongside my dislocation from nature, was my dislocation from my parents.

Not long before, I'd been living abroad, in Australia. And, after that, I spent another two years leapfrogging from one African country to the next, writing stories for a rather earnest financial magazine. And it was while working for this magazine, on one of my many work trips abroad, that I had met Mary, on one of her many work trips abroad, at the bar of a hotel in Lagos, Nigeria. Beautiful and intelligent, Yorkshire-bred but worldly-wise, she was way out of my league – serenely sipping her Star lager with a cerebral paperback in hand. She carried with her an indefinable air of unaffected quirkiness. She seemed so naturally herself.

After about three or four Star lagers I plucked up the courage to make myself known to her. Approaching random girls at foreign hotel bars was not exactly a habit of mine. And, truth be known, I knew Mary was going to be there. And she knew I was going to turn up at some point. We had friends in common, it turned out. Both of whom had been in touch earlier that day to unite the lonely travellers.

Once we'd been through the awkward, ever-so-polite ritual of two English people meeting abroad – the 'ah, so you're from's and 'oh, so you know's – we very quickly moved into a harmonious, easy-going rhythm: a steady beat of mutually agreeable conversation, laughter and (occasionally pretty robust) debate that has never really left us.

I quickly found, drinking at that tropical Lagos bar, so far from home, that Mary is a brain-teasing combination of endearing contradictions: as indecisive as she is deci-sive (she can close a complex financial deal in the blink of an eye, but don't ask her to choose a starter). She is as much arts and crafts as she is finance and economics – idiosyncratic and yet reassuringly mainstream – and as at home in a Canary Wharf boardroom (or Lagos bar) as she is in a down-at-heel Wiltshire pub. She is cleverly creative without any kind of pretension or affectation – a gifted interior designer and jewellery maker – and at the same time a successful city banker, but without an ounce of entitlement or pomposity.

But what I love most about Mary, and what I saw in her that night, is that she is deeply involved in, and enamoured with, the business of life in all its fibrous detail. She has an unquenchable curiosity about all things, bordering on bloody nosy. And beneath all the elegant sophistication and world-weary wisdom is a determinedly kind, inquisitive little girl with frizzy hair and 1980s NHS specs.

Later that night we met her colleague, Malcolm, a cheerfully middle-aged, pink-cheeked and portly banking lifer, with trenchant (and unrelated) opinions on God and high finance. Mary and I took great pleasure in gently teasing him throughout dinner – our eyes catching in frequent knowing glances as Malcolm pontificated on the Catholic Church and UK interest rate policy – until it was time for one last Star lager and bed.

Not a day has passed since that cheerful dinner, thirty-three floors above the streets of Victoria Island in Lagos, when Mary and I have not communicated in one way or another. We started out with quite polite emails, progressed to text messages, through telephone calls to drinks in my favourite dingy London pubs and overpriced West End restaurants. Weekends together, holidays together, houses together and, finally, married together. It all just felt so easy. I couldn't believe my luck.

As a consequence of all this falling in love and travelling about the globe, I was rarely at home. And I rarely visited Mum and Dad. Life was just too damn busy. I took for granted the firm foundations Mum had built for her family. The foundations that had enabled me to become the person I was.

No matter where in the world I found myself – how dangerous or dull, unfamiliar or daunting – I always had this feeling, embedded within me, of stability. No matter what life threw at me, no matter how anxiety-inducing or uncertain my day-to-day life was, I knew there was an immovable stake in the ground, buried in the Hampshire soil, that would hold me secure in good times and bad, and never shift. Until, of course, it did. And suddenly I was spinning off into the unknown. But as I listened to that lark it began to dawn on me, for the very first time, just how nature could help to pin me back down.

This song, the song of the skylark, is a sound you might well have heard many times before without ever realizing it's a lark. William Shakespeare said in Sonnet 29 that the lark *From sullen earth sings hymns at Heaven's gate.* But this is just a small window into its wonder, and the reverence humans have held the lark in for century upon century. In fact, in Shakespeare's time his fellow poets and bards – the beasts – would eat the tongues of larks in the vain hope that by doing so these tongues might somehow inject the ecstasy of a lark's song into their turgid seventeenth-century orations.

If at any time in your life you've been a country dweller, or spent time walking in open fields or over downland, then the skylark will have formed the musical backdrop to your walk. Next time you are pounding an open field, stop walking, cock your ear and listen. The best time to hear them is early morning or early evening in the spring and early summertime. And if you hear that soaring, tinkling, joyous sound, and you're still not sure it's a skylark, then you can make doubly sure by looking up in the air. Because skylarks are one of the few birds that sing while they're flying. If you look up and see a small streaky brown bird, with white tail flashes, hovering thirty feet or so above your head, then you've correctly identified the little fella. Undoubtedly there will be more than one lark as you strain your neck muscles. As you look up, you'll begin to notice more of them. Some will be hovering about, while others will fall and tumble across the sky, battling over territorial rights. That tumbling and falling flight of the skylark, and the joyous singing on the wing, is the inspiration for the phrase 'larking about'. Although I've not actually heard anyone use that phrase since my aged history teacher told me to stop 'larking about' around May 1988.

After I cannot remember how long immersed in the song of the lark, my brain sent a signal to the rest of my limp body that it was probably time I got going. As did my shoulders, back and bum, which by this time had become completely saturated. The light was fading a bit and Mary would be wondering where I'd got to. I stood up, shook myself down like a wet spaniel, turned my face into the rain and began to walk home. Suddenly the countryside around felt alive to me in a way that it had never really done before, or, at least, not for many years.

A hare cantered across the field perhaps fifty yards in front of me. I noticed the graceful ease with which it moved across the ground. Unlike a rabbit, which lurches and hops about, the hare slides across the landscape in what feels like a single elegant movement. To watch a hare move is to believe that Mother Nature is an artist and not a scientist. Hares are as beguiling as they are wondrous. And if you see a skylark in an open field, you are probably not far from a hare. Hares look and feel ancient. In a way that their cousin the rabbit just doesn't pull off. I've always felt that the difference between a rabbit and a hare is the difference between a Vauxhall Astra and a Bentley. And you can lose yourself in those hypnotic hare eyes.

Each summer, as a child, my brother and I were given the somewhat unenviable job of pulling out the wild oats that had taken root in the fields of barley that grew on my father's farm. Wild oats are the enemy of the barley farmer because they degrade the crop, and instead of selling your barley for lots of money to brewers to make beer, if it's full of wild oats, you can only sell it for a pittance for cattle feed. And so I was given this important job alongside my elder brother. And I loved it. Not only was I paid a staggering £1.50 an hour, but I was accompanied as I worked by the song of what sounded like a hundred larks and, as I paced the fields of barley gathering those rogue oats, hares would pop out from all angles.

One day, as I was plodding along methodically collecting oats, I saw the ears of barley seventy or so yards ahead of me shiver. This shiver suddenly started moving with great speed directly towards me. I was gripped. And before I had time to react, a large brown hare crashed with a great thud into my legs and lay, stunned, at my feet. I looked at the hare. And the hare looked at me. We were as shocked as each other, I think. And for a stunned second or two I peered into those beguiling brown eyes and something indefinable stirred inside my ten-year-old self. It was such a powerful feeling that I can still recall that moment of hare connection with utter clarity over thirty years later.

And yet for so many years these animals became forgotten friends as I charged on with my life: striving, striving, striving to succeed in work, in love, and find my place in the world.

Each step I took on my way back from the hillside brought a new wonder. Meadow pipits sprang from the ground with a high-pitched *pcheep-pcheeping* noise, and yellowhammers flashed before me in flourishes of brown and gold. How had I missed all this, and for so long? If the yellowhammer were a pudding, it would be a sticky toffee pudding, drizzled in custard and with generous dollops of golden syrup. It's a type of bunting. And I only mention that because I love the word 'bunting'. And I love that our ancestors evolved such beautiful names for our birds, like 'bunting' and 'pipit' and 'lark'. And 'peewit', too.

Peewits (otherwise known as lapwings or green plovers) are, in fact, a bird of the coast – a wader – but they breed up on hills during the spring and summer. And if you scan the sky in February, you might see great flocks of peewits circling up above looking for suitable places to nest. I almost drove off the side of the motorway when I saw just such a sight not that long ago (once you develop a love for birds, almost crashing cars will become a common occurrence, I'm afraid).

It is an arresting-looking creature, the peewit, and easy to spot in a crowd: deep green of back, white-and-black-breasted, and with a magnificent plume that carves up from its head in a perfectly formed arc, the beauty of which only nature could provide. It looks not entirely dissimilar to a 1980s New Romantic. And it's easy to spot in the air, too, because of its blunt wings – like someone's cut off the tips – and in spring they dive and cartwheel in the air, crying, 'Peewit, peewit!'

The peewit nests on the ground in wide-open spaces, as do the skylark and the meadow pipit. And these days, now that I am familiar with their habits, I always make sure when I am walking over the downs in the spring never to tread accidentally on one of their nests. These poor birds have got enough predators to deal with already – cats, badgers, foxes, stoats, weasels, crows and magpies (to name but a few) – without becoming the victim of my misplaced fat foot. The lapwing, though, has an excellent ruse to ward off these predators (and my fat foot). In the words of the late Will Meade, the Wiltshire dialect poet:

> Look at the peewit ther'
> Floppin about!
> Her be afeared, 'cause her
> Nest's hereabouts.
> Pretendin' her wing be broke,
> Artful o'd thing
> Walk on a bit – you'll see
> That cures her wing.

I don't know how good your Old Wiltshire Dialect is these days but, in summary, the female lapwing feigns injury to distract predators from her nearby nest. And this artful old trick is also why the collective noun for a group of lapwings is a 'deceit'. The stanza above comes from a longer poem by Will Meade about how he is never alone while tending to his sheep high up on the Wiltshire Downs. He talks of the starlings and jackdaws, lapwings, hares and rabbits he encounters as he tends his flock.

> Lonely here? No, not I,
> How can I be,
> When on our Plain be so
> Much company?

I quite agree. How could I ever have thought I could possibly be alone, as I lay down on the hillside? And I've not even covered a half of what I might have seen up on that Wiltshire hill: linnets, wheatears and stonechats, to name but a few other birds of the open country. And what I love, even more, is that there is still so much I don't know, years after my initial birdsong epiphany. I am regularly bamboozled by what I see and hear when out walking, despite years and years of looking and searching and finding out. All those different butterflies for a start! And I love that nature can do that to me. Since it began in earnest on that hillside on that life-changing day, my odyssey has never ended. And on every walk I do, I learn something new.

But the king of all the birds of the open country, to my mind, remains the skylark. Not much to look at I'll admit – brown and streaky – but with the voice of an angel. And a voice that's inspired more than one or two poets, composers and writers through the ages.

In fact, the skylark is a firm fixture in the canon of European folklore and verse. Since man first evolved to sing a song, tell a story or scrawl a poem, he has praised the power of the lark to move his soul. If you listen to only one piece of music today, make that *The Lark Ascending* by Ralph Vaughan Williams. Quite often I will find a suitably peaceful part of the garden, hopefully with the sun warming my face, lie down on the grass, close my eyes and press 'play'. Who needs meditation, breathing exercises and yoga when you've got Vaughan Williams and a deckchair? (If it's too cold outside or raining, I find instead a warm, quiet room and a comfortable armchair.)

And it's not just famous composers who've been

moved to pour their creative energy into lark love. The great romantic poets of the eighteenth and early nineteenth century weren't above it either. To be frank, though, the first thing that goes through my mind whenever I read the opening stanza of a poem by one of these old romantics, is: *Bugger me, this is going to take all afternoon to read.* I generally believe that when it comes to poetry (and speeches) brevity is the soul of wit. But every so often these romantic poets hit you square in the eyes with a line or phrase. And so it is with Percy Bysshe Shelley's poem 'To a Skylark':

> Hail to thee, blithe Spirit!
> Bird thou never wert,
> That from Heaven, or near it,
> Pourest thy full heart
> In profuse strains of unpremeditated art.

Strains of unpremeditated art. That's it, Percy. That's exactly it. He finishes up like this:

> Teach me half the gladness
> That thy brain must know,
> Such harmonious madness
> From my lips would flow
> The world should listen then, as I am listening now.

It's been said that this is not so much a poem about a skylark at all, really, but a search for something ideal and elusive, something that cannot, in the end, be captured in words. And as I lay on that hill after that difficult, life-altering day, staring into the leaden sky and with my skylark singing exultantly above my head, those lines are not far off how I felt. Who needs words when you've got a skylark's song?

Skylark

What it looks like: *About the size of an adult fist, the skylark is a streaked brown and white bird with a very distinctive Mohican on its head. In flight you will see they also have white tail flashes. Not to be confused with meadow pipits or corn buntings, which look similar and live and nest in the same wide-open spaces.*

What it sounds like: *An ecstatic series of notes that seem to ripple out across the air. They are one of the few birds that sing on the wing – they hover about forty or fifty feet off the ground above their territories – so if you hear a rolling, cascading song in a field or meadow, and you're not sure what it is, just look up in the air to confirm. Meadow pipits and corn buntings don't sing on the wing or hover like the lark.*

Where to find it: *Skylarks are a bird of the open country and farmland. Whenever you are out and about in a wide-open field, high downland or low arable fields, especially in spring and summer, you are very likely to hear and see a skylark. They make their nests in small depressions in the ground that are very hard to spot.*

What it eats: *Extremely varied. They'll eat worms, insects, spiders and slugs, as well as seeds, leaves and even some wild flowers.*

Chance of seeing one*: In a built-up area 0%. But if you are in wide-open arable farmland anywhere in Britain, then there is an 80% chance of seeing or hearing a lark, though numbers have fallen sharply in the last forty years due to intensive farming and overpredation by badgers, crows, foxes and others.

* This number is not at all scientific and is based only on my own vague wanderings around the towns, villages and wide-open spaces of the British Isles. I have applied it to each of the twelve birds in this book.

2. Robin

It is enough
To smell, to crumble the dark earth,
While the robin sings over again
Sad songs of Autumn mirth.

                              Edward Thomas

To see a World in a Grain of Sand
And a Heaven in a Wild Flower
Hold Infinity in the palm of your hand
And Eternity in an hour
A Robin Red breast in a Cage
Puts all Heaven in a Rage

                              William Blake

A turn or two I'll walk
To still my beating mind.
        William Shakespeare,
        The Tempest

Never was there a more ubiquitous, easily recognized and reassuring presence in our lives than the common robin. It is the national bird of Britain, decided via a poll by the BBC in 2015. And for good reason. You cannot walk more than a few yards in any town or village, park or garden, without clapping your eyes on a robin. And if you haven't seen the robin, you can be damn sure he's seen you. You're probably invading his territory, which he will deeply disapprove of, or you might be digging in the garden and disturb a juicy worm for his lunch.

And you could not find a more iconic or characterful creature to start your journey of bird rediscovery with than a robin. Not least because no other bird will ever get as close up to you as a robin will. And they'll even eat out of your hand if you offer them something tasty enough. They are as fearless as they are inquisitive.

In fact, I would defy anyone *not* to love a robin. This love, I suppose, is born out of the robin's eternal presence for us humans: easily identified, with that thick paint-smear of orangey red on their breasts, and never far away. And they've also always had a strong association in our minds with the season of goodwill. Robins have appeared on pretty much every Christmas card since Christmas cards were invented in the 1840s. Their link with this time of year was forged by the fact that the very first postmen wore bright red frock coats, and so were nicknamed robin redbreasts. It was a natural evolution that robins started appearing on the cards they delivered, especially at a time of year when robins are most noticeable amid the bare trees – warming our cockles with their company on those cold, short days.

Since time immemorial people have been taking pleasure in the company of robins and learning lessons from them. One of the most successful children's books of all time was *The History of the Robins* by Sarah Trimmer, which was in print continuously from 1796 right up to 1914. It detailed the exploits of four redbreast nestlings that taught children lessons in how to conduct their lives. Mrs Trimmer wrote that natural history, and her robins were 'replete with amusement and instruction'. I could not agree with her more. And 216 years after that line was written I too was diverted and instructed by a robin. I too learned lessons in how to conduct my life. It was a robin that kept me company in a drab hospital car park – amid all the high stress of my mother's illness – and taught me, truly, the benefits of taking a breath amid nature, even at the heart of a noisy city and in the eye of a personal storm.

About two weeks before I came across my robin, I woke up one morning to find that a fat friend had placed a grand piano on my chest, and then sat on it. I couldn't breathe. And I'd never experienced anything quite like this before. I reached over for Mary's reassuring hand to grip on to, but the bed was empty. The startling remembrance that Mary was away on a work trip only made matters worse. I felt immobile. Adhered to the sheets by my own sweat, and not enough oxygen in my lungs to power a mouse. I'm not usually one prone to hypochondria, but I really did, for the smallest of split seconds, think, *This is it, sunshine: curtains.*

Once I'd convinced myself that I might just limp across the line to my thirty-seventh birthday, I decided to do a bit of private investigation. But Dr Google is a dangerous friend. You tap in a few minor symptoms and, the next thing you know, you've got three different types of cancer and have suffered, or are on the verge of suffering, multiple organ failure. It reminded me of a favourite scene from Jerome K. Jerome's book *Three Men in a Boat*. Jerome is suffering from a *slight ailment . . . hay fever, I fancy it was*, and makes the mistake of looking up the symptoms in a medical dictionary. *I forget which was the first distemper I plunged into – some fearful, devastating scourge . . . and, before I had glanced half down the list of 'premonitory symptoms', it was borne in upon me that I had fairly got it. I sat for a while, frozen with horror; and then I came to typhoid fever – read the symptoms – discovered that I had typhoid fever, must have had it for months without knowing it . . . Cholera I had, with severe complications; and diphtheria I seemed to have been born with.*

He finds that he has every other disease in the book, of course. Except for *'housemaid's knee'*, which he feels rather hurt about. *'Why this invidious reservation?'* Dr Google is just the same. And it was a measure of my mental and physical state at that time that I convinced myself, like Jerome, that I was not long for this world.

For the previous month or so I had been trapped in a kind of cancerous twilight zone. I recognized the faces and places, but the landscape was fuzzy and unfamiliar. I found myself daily trying to force together an intractable puzzle, of which none of the bits were designed to fit. It was incredibly hard to disconnect my brain from caring for my dying mother one moment, and plug it seamlessly into grown-up, heavily earnest reports about the future of business and finance (my day job) the next. I ended up spending large parts of my work day loitering on Fleet Street smoking furiously, and talking endlessly to Katie and Richard on the phone about how we were going to solve our insoluble problem.

And yet I made it my absolute aim that I would not miss a single deadline at work. My reports would go to the printer like clockwork as usual every week. I would not let this private tragedy affect my public life. Strong people don't do that. They just get through.

Above all else was an all-consuming need to please everybody. I would never miss a call from family – no matter how busy I was at work – and if even the slightest news came from Dad about Mum's condition, I would drop whatever it was I was doing, jump in the car and tear down to the hospital from my office (100 miles away). And then I would sit at my desk till 10 p.m. to get the work done that I had missed, not because I was particularly fearful of what my boss would say, but because I did not want to put out the people who wrote and designed the reports that I edited. I could not let them down.

I became utterly exhausted by it all. And yet, infuriatingly, the more tired I became, the less I seemed able to sleep at night. At the end of the day I'd appear at home like a ghoul, shuffling about, opening cupboards and then closing them again, looking listlessly in the fridge – as if I might find an answer to my problems in the freezer compartment – or simply collapsing on to the sofa with a stupefied glaze. And as I lay there, thinking hard about nothing at all, Mary would walk in.

'I really don't think you're OK.'

'I'm absolutely fine, I promise you. Just a shitty day at work, that's all. I'll be back to normal again once I've had a drink.'

And then I'd go to the kitchen, lie on the table, and scream at the ceiling.

On other days I'd meet up with an old friend or two and get mind-bendingly drunk, before returning home in the small hours and passing out on the sofa listening to songs from the 1970s on my computer. I'd never had a particular penchant for seventies music, but in my drunken and reflective state, eager to escape to any-where, *any time at all but right now*, all I wanted to hear were sentimental songs from that decade: 'Seasons in the Sun', 'American Pie', 'Landslide', 'Those Were the Days', 'Stairway to Heaven' – over and over again. And, weirdly, more than any other song: 'Where Do You Go To (My Lovely)' by Peter Sarstedt.

I'd wake up in the morning, foggy, to find Mary had been down at some point to put a blanket over me, remove the headphones, and unplug me from my drunken attempts at time travel.

We'd only been married six months. In any normal marriage this would have been the honeymoon period: halcyon days before the onset of children, where you don't argue and go on carefree mini-breaks to glamor-ous European cities every weekend. That had been the plan anyway. That was what had been in the 'early mar-riage' brochure for the both of us. But instead of all that we got a glioblastoma (grade 4) on our marriage platter and a family in meltdown. I promised Mary all that 'for better' stuff, and all she'd got at the start of our marriage was a bucketload of 'for worse'.

Throughout all this I continued dutifully transmitting

cheerful messages to the outside world from our little cancer bunker, about how optimistic we all were and how strong Mum was. But the truth was the complete opposite. Mum's condition deteriorated from the moment she was diagnosed. The news never got any better. No matter how much I willed it to. I would take this bad news, turn it round in my hands like putty, remould it a bit, and transform it, against its will, into good news. I tried everything I could to deny the reality of my situation.

Mum had been through a punishing course of intensive radiotherapy and chemotherapy. Enough, I would think, to sink a small oil tanker. And, as Richard put it, she was on more drugs than Keith Richards on tour.

I turned up at home one soggy-leafed and grey-skied day when Mum was recovering from her latest batch of radio and chemo. She was lying on the bed, propped up on a couple of cushions, watching television, and with her adoring little black spaniel, Quink, positioned over her lower legs. A spaniel that had never left her side. Not once. Except when my father physically put her outside to do her business, this wonderful dog – our little canine Florence Nightingale – did not leave my mother's sick bed. She made Lassie look heartless. And it put into all our minds a poem by Rudyard Kipling, which, a few months later, was read out at Mum's funeral.

> Day after day, the whole day through—
> Wherever my road inclined—
> Four-Feet said, 'I am coming with you!'
> And trotted along behind.

Now I must go by some other round,—
Which I shall never find—
Somewhere that does not carry the sound
Of Four-Feet trotting behind.

Over the previous six weeks Mum had been through the mill, and back again, and once again for good measure. The problem with curing cancer, or at least attempting to control cancer in this case, is that the cure is much worse than the disease. It's a bit like using a machine gun to extract a splinter. The splinter will disappear all right, but so will your arm. It was the cure that ultimately did for our mother. Severely weakened as she was, not just by the chemo and radio, but by a brain operation some weeks earlier, she had what was later described to us by the doctor in cold institutional terms as 'an episode'. She began quite regularly to have these 'episodes'. And it was on this particular soggy and grey afternoon that she was struck down with one and nobody was at home but me. One minute she was awake and coherent, and the next minute shaking uncontrollably and gripped by a vicious fever.

I rushed off to the next room and dialled for an ambulance. And went back to attend to Mum. She was in a real state. The most disconcerting thing was that it felt like it was a different person lying on the bed to the one I'd been happily chatting to just minutes before. It's amazing how sickness can make complete strangers of people. In those dramatic moments, as her body fought to keep alive, my familiar old mum went away somewhere. And I was terrified she would never return.

And where the hell was Dad, for God's sake? I repeatedly tried to call him, but it kept going straight to answerphone. Even more horribly, Mum had recorded Dad's answerphone message herself, because he is technologically backwards, and so it was her cheerful voice at the end of the line every time I tried to call. 'This is Peter's telephone – please leave a message.' She had one of those voices that you could listen to on the telephone and literally hear the smile. And yet as she spoke these words, smiling down the phone at me, there she was, unrecognizable on the bed in front of me.

Oh, dear God, where is that ambulance? My mind was racing. I felt utterly helpless. Nobody I called would pick up the bloody phone, and there was absolutely no sign of the ambulance. Eventually I got through to someone, Richard, I think, and he said he would be with me as fast as he could. That at least was some solace. And then, out of the window, I saw that blessed, wonderful ambulance blinking its way into the driveway. But then it stopped. Horror. It reversed the way it had come. And left again. My heart collapsed. I could not comprehend. I called the hospital, incredulous. When finally I managed to get through to someone who could help, they told me it was most likely that the ambulance had been called off on a more urgent case. A more urgent case? What could be more urgent that this? I hated myself. I should have been more forceful when I had made my original call. I should have been clearer that this was a desperate case in need of immediate and urgent assistance. Instead I had become a victim of my own upbringing. Of that ingrained fear of not wanting to make too much of a fuss. The one time in my life when I really should have made the fuss of all fusses – shouted forcefully down the wire and demanded immediate help – I had dithered. Irony of bitter ironies, it was from Mum I probably inherited the 'don't make too much of a fuss' gene.

It was another two hours before the ambulance returned. And Mum had deteriorated to a point where she needed very serious and very urgent care. It was the beginning of the end. And I have never really forgiven myself for not doing more – for not chasing that damned departing ambulance down the road and banging on the doors till they came back. And to hell with anyone else. Mum was later admitted to Ward 4C of Southampton General Hospital (a name forever embedded in our family's consciousness). She never spent another night at home.

I became consumed by guilt. It's the guilt that is the engine room of stress when someone close to you is suffering. It's the guilt that means you spend days, and nights, at the hospital. It's the guilt that steals your sleep and bullies your conscience into submission. It's the guilt that becomes your master.

In the weeks and months that followed we made sure Mum was *never* alone in the high-dependency unit. Not even at night. Dad was at her bedside each day without fail. Katie would 'commute' from Edinburgh two to three times a week (a mere 1,000-mile round trip) while simultaneously caring for her young children.

And Richard, ever haunting the corridors of Ward

4C, would bribe the nurses with fresh bacon and sausages from his father-in-law's farm, just to make sure Mum received the necessary level of attention. We even bought super-sized jars of coffee and enormous bags of tea for the nursing staff in a desperate bid to butter them up. It was mad. But that's what you do, isn't it? Everything. You. Can. But it never felt like enough. I was a helpless spectator.

I felt I was falling short of my responsibilities. I blamed myself for not being a better son. Not just now, at Mum's time of intense need, but all through my childhood and early adulthood: my badly behaved school days, sulky adolescent teenage years and absent twenties when all I had to do was call her and tell her a bit about my life but rarely did. And now it was too late.

What I should have being doing, of course, was getting decent rest and exercise, so I could have been of genuine help to Mum. But what I actually did was to hammer myself till I could go no further. I could not allow anyone else to see that this was affecting me. After all, I would think constantly, my grandfather – and so many others of his generation – had lived through two world wars, the Depression and all the countless other private and public tragedies that made up the twentieth century – but they had got through. And I needed to emulate them. Their ghosts lived constantly on my shoulder, whispering down to me through the ages. I could not show weakness. Above all else I needed to be strong for Mum. And then, inevitably, I found myself on my back at 5 a.m., with a grand piano

41

on my chest and wondering what hymns to have at my own funeral.

The days and nights melded into one. I lost track of time and place. And the road between my house and the hospital was so ingrained on my frontal cortex I could have driven it blindfolded, with one hand tied behind my back and drinking a gin and tonic. I'd even developed a taste for the hospital's chemical coffee, though I never did quite get to the bottom of how they managed to make it stay hotter than lava for such a sustained period of time.

I was wandering the hospital grounds one day, some-what dazed after a particularly gruelling few hours in Ward 4C, when I stumbled across a robin. He lived in Chalybeate Close this robin, a cul-de-sac that hosted a private medical clinic not five minutes' walk from the hospital. It was a place I often took refuge at odd times of the day, when things became a bit much. It was a small haven for me – a not particularly attractive concrete and red-brick haven, I'll admit – but a place where I could walk and think and smoke an occasional illicit cigarette. There was also a thick clump of unruly woodland in one corner. And this is where my robin sat, bold as brass, observing me – inquisitive little thing that he was.

I can't pretend to you that he worked any particular miracle for me that day. Except to watch me. As I watched him. Smoking my cigarette. But I made sure to look out for him every time I went back to Chalybeate Close. As the days passed he became a friend and helpmeet: reliable and unruffled. Disinterested in my plight but a reassuring presence: like the old-fashioned policeman at the end of the street. This little bird became a calming influence on my mind – a positive and life-affirming distraction – and so much more effective than my carbon monoxide-filled Silk Cut.

And if I watched long enough, other birds would make themselves known to me in Chalybeate Close, this rather drab, innocuous corner of a southern English town: blue tits, great tits, an occasional blackbird and, if I was particularly lucky, the odd charm of goldfinches would tinkle past. Goldfinches possess a beauty that is striking – and unmistakable – with their red, gold and black livery, and gentle tinkling chirrups. In days gone by, in some places, they were known as thistle finches (because that is what they love to eat) and, more appropriately to my mind, King Harry Redcaps. Such was the popularity of 'King Harry' back in Victorian times, that bird catchers would roam the country capturing them to sell as caged birds. By the 1860s they were almost extinct, in fact. I am very glad to say that this practice was banned long ago, and now goldfinches can be found tinkling in great numbers in town and country alike.

But it was my robin that lay at the heart of this joyous crowd of little birds. The gatekeeper. It was his presence that first hinted to me that it might pay dividends to stop and watch and listen, instead of crushing my cigarette end and rushing headlong back into the Ward 4C melee.

On one of my visits to Chalybeate Close I managed to get so close to my robin that I could see his chest feathers vibrating as he sang. He didn't seem to mind my presence so near to him at all. As the robin sang and his little orange chest heaved with the delicate strains of melody – I could even see each individual feather moving – all my stress fell away. Ward 4C, my hundred-mile-an-hour life juggling work, family and friends, all that irrational guilt, even Mum and the cancer, retreated timidly into the background – and, just for that moment, my beating mind was stilled. I found peace.

Robins are one of the few birds that sing all year round. Most birds sing to protect their territory only when they are breeding in the spring, but so protective are robins of their realms that they bullishly announce their presence through all the seasons. Unlike most birds, male and female robins look exactly the same and in the winter months both will sing. People have called the robin's song metallic and melancholic, which kind of felt appropriate during my times in Chalybeate Close.

Robins nest all around us, too. Entire theses have been dedicated to the profusion of extraordinary places that people have found them nesting. They seem to nest anywhere but in a tree: in upturned paint trays in sheds, a disused beehive, old hats and even – in one case – on someone's bed.

I felt it to be such an enormous privilege to get so close to this wild little creature: an indefinably beautiful wonder of nature – small, perfectly proportioned and fierce. I love robins and other wild creatures simply because they *are* wild and unsentimental. They don't ask you questions and they don't answer back. And some-times in life that is just what we need.

So inspired was I by this feeling of stillness the robin stirred in me that I decided to invest in a series of bird feeders. I wanted to bring a bit of that Chalybeate Close calm home with me. Previously I had regarded bird feeders as uniquely the realm of old ladies and lonely bachelors, but it was a transformative purchase for me. Establishing my small colony of bird feeders was a bit like going on a course of stress-busting medication, except in place of pills I got birds. Once my feeders were up and had started attracting visitors, it became a daily obsession: blue tits, great tits (full names: blue titmouse and great titmouse), sparrows, nuthatches, chaffinches, greenfinches, goldfinches, blackbirds, wrens, the occasional siskin, robins (of course) and many more. A daily reminder of something I had long ago forgotten – the sheer diversity of birdlife we are privileged to be surrounded by. I could watch them for hours. It was the first thing I would do each morning – inspect the feeders from the comfort of my warm kitchen. In fact, one particular neighbour, whenever he saw Mary, would (rather unfairly in my opinion) ask her the same question: 'How's Charlie? Still watching his birds?'

It was once said of the poet T. S. Eliot that he could hear the grass grow. I've always loved that description. And it puts into my mind the best advice I've ever received about getting to understand, and to love, nature. I was told, simply, to find the time to sit still. I could not remember the last time I had settled myself in the shade of a tree and just sat and watched the world around me – immersed myself into the landscape. And not just for a hastily grasped five or ten minutes, but a sustained period of time. An hour even. Just doing nothing amid nature.

When I was younger I had often read books by the great twentieth-century writer and naturalist Denys Watkins-Pitchford. He wrote about the daily happenings of nature as he walked the Northamptonshire countryside with his dog and his gun. On the flyleaf of every book he would quote an inscription his father once found on a Cumbrian gravestone: *The wonder of the world, the beauty and the power, the shapes of things, their colours, lights and shades; these I saw. Look ye also while life lasts.*

I became utterly determined to find the kind of peace that would let me live up to those words, which I realized even then was a bit of an oxymoron. It sounds so easy. And it is very easy and profound to write about the 'sit-still-and-watch' approach to life. To talk about poets listening to grass grow and advise everyone to be at peace and commune with nature, and to 'look ye also'. But the reality, I found, was very different. In actual fact when I embarked on my project of sitting alone under trees amid nature, I found that being alone with my thoughts was not only very daunting and somewhat scary, but also not at all peaceful.

My instinct was to run away from the silent moments, afraid of what I might find there (or drink my way through them). And so many of my attempts just to sit, watch and listen to the grass grow would fail utterly. My phone, that digital siren in my pocket, would summon me – a message from work I simply could not ignore or a sweet text from a friend that I felt if I didn't answer right now would look rude. Other times, as I stared into space, some dark thought would steal my peace like a thief in the night.

It took me quite some time to master being alone amid nature. To be still. I still fail at times. But when I do succeed it makes all the difference. And nothing else really matters. I pick up little points of interest that I never would by reading books or watching documentaries; it becomes personal, up close like that, and the birds become part of my daily story.

Watching my bird feeder gave me an insight into all the different hierarches of the birds – quite literally the pecking order of their lives. I found the sparrows to be the most dominant species on the feeder, with the smaller great tits and chaffinches waiting patiently in the background for their fill. At the bottom of the hierarchy were the delicate blue tits and coal tits – nature's pretty little Oliver Twists. And let's not forget the humble dunnock – a somewhat shy and self-effacing bird, in its streaky brown and grey uniform. But they are not to be underestimated. The female dunnock will always mate with two male birds so that neither knows who is the father, and so both will tend to Mrs Dunnock and her chicks during the breeding season. Clever.

Different birds would come to my feeder at different times: a family of nuthatches would appear, like clockwork, around 9 a.m., followed by a couple of greenfinches and sometimes a siskin. And the black-and-white great spotted woodpecker – a giant among the tits and finches – would send them all scattering in a flurry of panicked flutters. The big bully. I also worked out that if I put niger seed in one of the feeders then I would always – without fail – attract goldfinches. I could summon goldfinches at will! Just the thought of it gave me a warm glow.

But it wasn't just about Chalybeate Close and the bird feeders. In the days and weeks after the Attack of the Fat Piano Man, I needed more. And, despite all my researching, I never did quite get to the bottom of what had happened to me that disconcerting morning. Someone suggested to me it might have been a mild attack of stress-related angina. But whatever it was it had frightened me, and I resolved to create time for myself. To shoehorn hours into the day where I could just get out and walk about a bit in a wild place. To rest under an old oak for an hour or so. If I was lucky enough to find time to get off the beaten track, then I would put my guilt carefully on a shelf, leave the phone at home and take myself off to a place where I knew there would be wild things. Where I knew that if I just waited, I would not need to find nature, because nature would find *me*.

It was a long road but I had started in earnest. I began to see how I might one day be able to walk in Mr Eliot's shoes and listen to that grass grow. It helped that as my understanding of the power of reconnecting to the wildlife around me began to evolve so did the season. It was autumn. And the thing about autumn is that it has this inbuilt calm about it. The mad stress of high summer and holidays has passed and the world settles back into its routine. Nature begins its gentle decline into winter with a happy sigh. Those viciously hot and thundery days of August are behind us and we've not yet moved into the stormy dawn of winter. Everyone, nature included, takes a breath. And so did I.

For the first time in many years I became aware of the changes that take place in nature in autumn. They are profound, at times inspiring, and too often ignored. It is a time when our avian friends are on the move all across the world. When vast swathes of geese blacken the skies as they journey down from the north. When every estuary and inlet is infested with bearded men in beanie hats and high-powered binoculars, their eyes peeled for rare migrants. They are right to be excited. There is an ancient mystique to bird migration that can be intensely moving, and it's not only the summer migrants returning from Africa in the spring. The skies also come alive from September onwards, as the birds of the northern part of the planet move south for warmer winters.

And if you are lucky enough to experience one of these migrations, it can make the heart miss a beat. One cool November evening a few years ago I did just that. I was walking down to the sea with Mary and her mother in the gloaming on our way to a beachside memorial for Mary's grandmother. A contemplative silence had fallen upon us. And then we heard the characteristic gentle honk, honk, honking of approaching geese. All of a sudden the empty reddening sky was black with an airborne cavalcade of brent geese on the final leg of a 3,500-mile journey from Siberia. Wave upon wave of these glorious creatures sliced through the air above us. And they just kept coming. It left us without the ability to speak. Our necks cricked and mouths hung open as we admired this marvel of nature for thirty happy seconds. I'd never seen anything like it. And then, when the final V-shaped wave had soared past, my mother-in-law turned to Mary and said, 'It was very near this spot forty years ago that your late father proposed to me. The brent geese flew past that evening too.' We all felt very warm inside.

The walks I took during Mum's illness became utterly integral to relieving my stress – and stilling my beating mind. But it wasn't just the sights that brought me solace, it was the smells too: smells that fired my senses and transported me, arrow-like, away from the pain-filled present and into happier, stress-free times. I know it's become a cliché, but John Keats had it spot on when he described autumn as a time of 'mists and mellow fruitfulness'.

These autumnal walks resurrected in me tranquil memories of childhood expeditions foraging the hedge-rows for blackberries and hazelnuts. In fact, it was one of the last activities I undertook with Mum before she became too ill to walk. We wandered together, with Mary and the dogs, to all the old places around the farm we'd visited as children, in search of blackberries for Sunday's blackberry and apple crumble. The deep, rich peppery aroma of rotting leaves reminded me of languid Sun-days that smelled of woodsmoke and roast beef, of Dad in his tatty old oil-stained jacket on his Ferguson tractor, a battered old trailer full of apples (and Katie and Richard) bumping along in its wake. It awoke in me the simple thrill of finding a conker tree in a friend's garden, or the warm light glowing from the windows of our house as I arrived back from school with Mum and the nights drew in on us. The gentle daily hustle and bustle of a life I once knew. This season and those birds were a constant reminder that I might yet have that life again one day. But I had to tune in to find that out.

As the shadows lengthened and the air became crisp, all of nature was bathed in a kind of weak sepia light. Wrens and chiffchaffs knocked out some late-season autumnal melodies, and the swallows and house martins formed great mobs, dancing through the sky and sagging down the telegraph wires, readying themselves for their long journey south again. I thought, *This is how I imagine heaven might be at certain times of the year. And if it isn't, I don't want to go.*

I began to take every opportunity I could to meander down leaf-strewn lanes, abundant with the fruits of the field, hedgerows bustling with life. And I found I could do so without any guilt. My mind was focused entirely on the changing season around me, and the birds. At weekends Mary would join me. And we'd walk together in companionable harmony, soaking it all up.

Blackbirds would squawk testily at our approach and then disappear with a petulant flourish into thick briar. Mobs of long-tailed tits would work their way along the tops of hedgerows in tandem with Mary and me – a humble wee goldcrest (the tiniest of all the songbirds) trailing in their wake. And sceptical robins would eye us suspiciously from neighbouring branches. Like reading a poem that moves the soul, they gave me pause, and allowed me on some deeper level to take stock. Or as John Keats once said:

> The redbreast whistles from a garden croft,
> And gathering swallows twitter in the skies.

This is the sound of the universe, I realized then. It's free. And it's on my doorstep. And no matter how stressful my life became after that – and, like the season around me, I was about to move into a much darker and wintry period of my life – I have never again been visited by that fat piano man on my chest.

Robin

What it looks like: *Both males and females look exactly the same, with an orangey-red head, throat and breast, and with a bright white belly and browny-buff back.*

What it sounds like: *The robin's song has a kind of metallic trill to it, and has often been described as quite melancholic or plaintive. While like most birds it is noisiest in the spring, the robin is one of the few birds to sing all year round, and both sexes sing in winter, which makes it easy to pick out at that time of year.*

Where to find it: *Robins are a ubiquitous presence in parks and gardens, in town and country, and because they are so competitive about territories you'll usually find that the robin locates you, and not vice versa. Robins are prolific breeders, having up to four broods a year, and will nest almost anywhere – although they have one of the shortest lifespans of all the songbirds at just thirteen months. Their chief persecutors, as with all songbirds, are domestic cats, which kill many millions of birds a year, and sparrowhawks – a devastatingly effective medium-sized hawk that hunts in confined places like gardens.*

What it eats: *Seeds, fruit, insects and weeds. But its favourite is a juicy worm, which is why it will always turn up whenever you start digging in the garden.*

Chance of seeing one: *If in a park or garden, town or country, then there is a 95% chance of seeing a robin.*

3. Wren

'Where can we hide in fair weather, we orphans of the storm?'

Evelyn Waugh

He who shall hurt the little Wren
Shall never be belovd by Men

William Blake

It was as I was staring zombie-like out of the window of (what passed for) my home office that I first noticed them. There'd been a pretty hard frost overnight and I was sitting at my desk, hands clasped for warmth round a steaming mug of hot, sugary tea, and doing my best to put off the deadline I had to meet that day. And then they began to emerge, the wrens, one by one, from the drystone wall ahead of me. The first had poked his head out rather furtively – no doubt checking the lay of the land – then another emerged, and another, and another. Surely no more? And then a fifth gingerly made his way out into the open, and a sixth. It was like one of those old-school magicians pulling rabbit after rabbit out of the same medium-sized top hat. I simply could not understand how all these wrens had managed to fit into such a tiny crack in the side of my wall.

It defied physics. And it defied all I had assumed before about the territorial nature of these quite fierce wee birds.

I shut down the piece of work I had limply embarked upon that morning and started researching wrens. Let's face it, far more important than a pressing deadline. I'd been watching the wrens all summer and autumn, taking immense pleasure in their ecstatic song, but I'd never really associated them with winter. This sight, and my subsequent research, changed all that. Despite being hugely competitive little blighters in spring and summer, fighting energetically over nesting sites with a winner-takes-it-all intensity, in winter wrens conduct a kind of Christmas truce. The fighting ceases and communal living begins. Because they are so tiny, wrens get particularly hard hit during wintertime and if it's a cold one, a quarter of them can die.

It is crucial at this time that they make up their differences and come together in small places in order to conserve heat. These little roosts often contain up to ten snuggling wrens, but the biggest roost ever recorded in Britain was a startling sixty-one wrens in a nesting box in Norfolk in 1969. Now, whenever I am out and about in winter, I will always keep an eye out for potential wren hideaways. It might be a little hole in a tree, under the eaves of a shed or indeed my garden wall.

The wren is one of the smallest songbirds but with the most beautiful song – a euphoric rush of rising and tumbling notes that spray out of its tiny breast and into the grateful air. It is the Dolly Parton of the bird world:

small, gutsy and with a voice that can be heard in three counties. In fact, it is almost impossible to conceive of how such a tiny creature can emit such a huge noise. I also wonder constantly how on earth these little brown feathery golf balls – with their stubborn, perpendicular tails – survive a harsh winter? Perseverance and grit, I always think, and coming together.

Like the wrens, we had come together as a family in that dismal winter of Mum's discontent. We'd had no choice. Often prone to the usual familial bickering and the occasional full-blown shout-as-loud-as-you-can row, we'd put our differences aside and hunkered down.

Growing up I can remember being a part of a tight-knit family but also somewhat apart from it. I am much younger than my siblings: five years younger than Richard and seven years younger than Katie. Rightly or wrongly that age gap always gave me a feeling of solitude within the family – a spectator and not a true participant. Not least because by the time I reached the age of five or six, and was looking for playmates, my siblings were all away at boarding school. It was in some ways like being an only child.

I used to yearn for their return in the holidays. And when finally they did get home from school it very often felt like I was trailing in their wake. I would spend very large amounts of my time in the back of a horsebox, reading Tintins, while my sister rode her pony in the various gymkhanas that took place in the surrounding district. Either that or playing cricket in the garden with Richard: bowling countless balls at him so he could hit them for six runs every time. 'I'm your elder brother,' he would remind me constantly (and still does). As though I might forget my place in the pecking order.

But the age gap also had its advantages. For one, Mum's legendary discipline was eased. While she was never far away from a smile that could light up your soul, she was also a tough old bird who didn't suffer fools gladly. There were lines. She owned two beautiful grey horses called Law and Order, which probably gives some small insight into her view on the world. 'I rather gave up when it came to you, Charlie,' she told me once in later life. Or in the words of either of my siblings: 'You got away with murder.' And I was regularly told: 'Mum would *never* have let me [insert crime or happy activity here] get away with that!'

Except for bad manners. We none of us ever got away with bad manners. You could burn down the French block and make a pyre out of *Tricolore* textbooks and she'd barely look up from her crossword. But if she found out you'd been rude to the fireman who put it out, all hell would break loose.

But despite occasional boredom and my lowly position in the family pecking order, Richard and Katie always behaved with deep kindness and affection towards their much younger brother. I was known as 'the brat' and generally treated as an object of amusement – and teased and doted on – rather than being seen as a direct competitor. They did, however, compete quite rigorously with each other. I often found myself acting as peacemaker. While they were, and remain, incredibly close – and one would do anything to help the other – they're also very different. Richard: whimsical, prone to introspection and hiding his angst behind a barricade of wry humour and quick wit. Katie: determined, life-bright and achingly compassionate – with a good dose of pragmatism thrown in. In good times these personalities were complementary, but at other times were powder keg and spark.

The vast majority of family rows, between any combination of the five of us, not just Richard and Katie, would happen during Sunday lunch – much to our mother's utter dismay. Anything might set us off. Quite often it would be Richard (always the joker) teasing my sister or me about some aspect of our lives – and Katie or me having a sense of humour failure about it. For example, when I put on a bit of weight in my teens and was a bit sensitive about it, he took great pleasure in calling me Beverly Buttocks, or Beverly for short. It infuriated me. And he was forever playing practical jokes on us all (and still does today at almost fifty years of age). Before you knew it, we'd all be at it – at full volume – and with Dad in the background, putting in his twopence worth – usually on an entirely different topic to the one we'd been so energetically arguing about.

'Who's taken my bloody wellingtons? Dick, you thieving sod! Why don't you EVER –'

And then suddenly, amid all the heat and noise, Mum would cry out in exasperation, 'PLEASE! All of you. Stop fighting! Why can't we just be like a *normal* family and sit down happily and eat lunch together?'

Silence. Then: 'Mum, just what exactly *is* a *normal*

family? Who are all these normal people you keep talking about?'

We'd all laugh. Just as quickly as the volcano had erupted, so it would die down again and we'd gather round the table in our little dining room and behold the delicious roast beef (or lamb, or pork, or pheasant) that Mum had been preparing for us all morning.

And then, like clockwork:

'Where's your father? Oh, for God's sake. He's done another disappearing act. And *always* when we're about to sit down and eat. PETER!'

Dad would usually have found a very important job to do at this point, that could not be done at any other time; no doubt he'd be double-checking the gate to the barn was securely shut and, while he was there, that there was enough hay for the donkeys – oh, and he might just fill up the sheep's water while he was up there. And where are those bloody dogs? (My brother had a Jack Russell terrier called Fly, and my sister had a beagle called Huckleberry Hound that would permanently be buggering off.)

And as Mum rushed outside to search for Dad, as he searched for the dogs, all the children she had so assiduously gathered for lunch and stopped from rowing would silently drift away again. Katie to the stables to check on her pony, Richard to the sitting room to check to see if the cowboy film had started (he wept for a week when John Wayne died), and me to my bedroom to inspect my Lego fire station.

Dad would eventually be located and dragged in, the

errant dogs trotting along happily beside him. He would take his place next to the joint, ready to carve. All together at last. Finally.

And then the phone would ring.

Dad would turn to Mum and say exultantly, 'That'll be your mother! She always calls just as we sit down for lunch.'

Mum (slightly red in the face now and through clenched jaw): 'No, she doesn't. It'll probably be your bloody father, and we'd be finished already, *like a normal family*, if you hadn't all disappeared just when the food was ready.'

Picks up phone in kitchen.

'Oh, hello, Mum. We're just sitting down for lunch.'

And Dad would call out every time, 'Oh, for God's sake, are we eating this side of bloody Christmas?!'

Despite these somewhat fretful lunches, there was a profound love at the core of my family. It cemented our little nest together and, in a funny sort of way, gave us permission for all the bickering. Because we always knew that, when push came to shove, we would come together again – in good times and bad. If ever I'm feeling a bit low and need a quick lift, I'll call my brother and laugh till my sides hurt (provided he doesn't call me Beverly). And if I want sage advice about life or emotional support in difficult times, I'll phone my sister. I'm incredibly lucky like that. All those arguments ever did was bind us closer, rather than push us apart. I never feel I know anyone properly until I've had at least one falling-out. If ever there were some external threat to any of us, we would flock together to protect one another. No argument any of us ever had was bigger than the family.

And it was those wrens that got me to thinking about the nature of our once warm and snug little home. The home that Mum had painstakingly built for us over four decades of marriage. I was lucky to have lived in the same house all my childhood. And the idea of it being sold, or someone else living in it one day, had always filled me with an unnamed horror. And seeing those wrens had given me a powerful urge to revisit those memories again.

All I wanted, right at that moment, was to be back home. Back amid the rooms, buildings, stables and barns that had so cloistered me as a boy. It was not far for me to drive and so I reasoned that I could probably pick up the reins of my pressing deadline later that afternoon. After all, I was properly distracted now.

I went into the kitchen through our old back door, always painted a thick dark green, and the very same door that would have stood there since the house was built. I noticed for the first time in a while the ancient and dented brass handle, polished shining bright by two centuries or more of warm hands clasping it. I have always loved our back door. At some point the Georgians had added a heavy cast-iron knocker in the shape of a fist clutching a wreath. I knocked on it, even though I knew nobody would answer. I just couldn't help it.

On entering the kitchen I found myself in a deserted nest. And everywhere I looked I felt this profound sense of absence. The cold silence of it all after so many years of heart-warming noise, delicious smells, and general family kerfuffle, overwhelmed my senses. No laughter or shouting, no eruption of barking dogs when I opened the door, followed by more shouting – 'Get down, get down, be quiet! Bloody dogs! Get down. Sorry, darling, I'm sure the mud will come off those trousers when it's dried.' – just cold, hard, silence.

I wandered into the dining room: scene of all those gloriously noisy and hilarious family lunches and dinners: Christmas, Easter, birthdays and anniversaries, winter, spring, summer and autumn. But it was just so many empty chairs now round an empty table.

Why can't we just be like a normal family? echoed around my head.

I'm so glad we weren't, Mum.

Nothing had changed physically – the fixtures and fittings all looked the same – nor had some of the smells – wet dogs, sheep shit and saddle soap – but something indefinable had gone. It felt like a totally different house. Bereft. All that remained, as I plodded around the empty kitchen, were memories. Colour-faded. Warm smiles of welcome, gone. All that indispensible love, dispensed. It was like seeing my home through the other side of the looking glass. Everything looked the same but was entirely different. If houses go through seasons, then this was the depth of winter for my home. Mum was unlikely ever to return to the soft-lined nest she had created.

Christ alive, what will happen to us all? I thought.

The architect and defender of the family, she had bound our disparate characters together and given our lives meaning. It turned out I could not rely on these empty rooms and buildings and barns for solace. It was home, of course, but it no longer felt like it.

I drove out of the gates and for the first time in years noticed the ancient oak at the end of our drive. It must have stood there for 350 years or more this oak. And while it gives the impression it's barely got any more life to give, it'll no doubt outlast me and possibly my children and grandchildren, too. It's so old that the centre of its trunk has become hollowed out. As children we'd climb up inside the tree and pop out on to one of its big boughs and survey our little empire. Fly, the Jack Russell terrier, would sometimes scrabble up after us, but she was more interested in the mice that nested in the trunk of this glorious life-affirming ecosystem.

Oaks are home to over 2,000 organisms. And they give off this particular pungently sweet aroma, too, in autumn and winter when the acorns fall. Whenever I smell that smell, wherever in the world I am, it transports me back to the oak tree at the end of our garden.

It was perhaps the single most photographed object in our garden. The pride Mum had in our home was reflected in the multitude of photographs she took of it, and us, during the course of her life. If I look again through the old albums, there is our house framed by this oak in sunshine, leaf fall or unexpected snow. When the snowdrops emerged in January, the primroses and daffodils in March, or the bluebells in April, there was the oak in the background. Year after year the same photograph, taken from the same angle. And the same excitement in Mum's heart at the beauty of her garden as it moved through the seasons. I realized then that this oak stood for everything that was solid and dependable about the childhood home Mum had built for us. And how much it meant to her.

That old oak, with all its sweet-smelling memories, loomed large in my mind as I drove back. When I got home I took one look at my computer, thought about my deadline for the smallest of split seconds, and thought, *sod this*. I resolved instead to head up to an ancient outcrop of beech trees I knew, which perched at the crest of a hill nearby. Like the old oak at the end of my parents' drive, these magnificent beeches had stood on the same spot for many hundreds of years, towering sentinels watching over the vale below. This vale was my home now.

It was a crisp, cold winter afternoon. The sky was impossibly blue, with an overblown orangey sun filtering weakly through the trees, showing up their beauteous naked shapes in plain relief, and making the world look like it'd gone black and white. My breath poured out of me like dry ice at a teenage disco. Beyond this high beech outcrop I could see perhaps half of Wiltshire spread out before me. A mist lay low under me, creeping round the feet of the hills and making their tops seem like islands in a sea of cloud.

It is an ancient landscape, the Vale of Pewsey. This sheltered vale has been a home for people going back 5,000 years and more. The hills around are littered with prehistoric burial mounds and Neolithic hill forts. You can't go fifty yards without tripping over a knapped flint or Bronze Age arrowhead. Not far away from where I was standing that day are the standing stones of Avebury and Stonehenge and, closer still, Silbury Hill, an ancient man-made enigma that no scientist or historian has yet to get to the bottom of. It is the land of King Arthur and King Alfred: the ancient county of Wessex. Legend and folklore ooze out of the very ground.

Up on that hill I felt part of something way beyond my limited understanding and short tenure on this planet. I tried hard to conceive of the people and animals that had lived before me in this ancient landscape. I wondered to myself just how many thousands of people must have made this beautiful place their home over the preceding centuries? And yet it felt so untouched by human hand, those endless, empty rolling downs and meandering chalk streams.

I looked out across the bare winter landscape and thought about how hard it is for the birds at this time of year, particularly the little wrens. Nature's bounteous autumnal larder – hedgerows that were once rich with berries and nuts – tends to run bare by December. And from then on – and up until spring – birds enter what is called the 'hunger gap'. Food becomes harder and harder to find. Skeletal hedgerows offer little solace. And the increasing frosts can leave the ground too hard for a hungry beak to penetrate. In particularly cold winters a great many of the smaller birds will die of starvation and the cold – like the pretty little wren.

And that is why they come together in winter – not just the wrens but all the birds – in large life-affirming flocks. Sometimes it will be one particular species that will form a flock, like starlings. Or, in the sunflower fields near my home this year, many hundreds of charming goldfinches, which would explode deliriously into the air around me as I walked.

At other times while walking over a winter field you might come across many different species, together forming a vast homogenous brown wave: skylarks, sparrows, chaffinches, linnets, thrushes, rooks and jackdaws (to name but a few) – as well as fieldfares and redwings, types of thrush that migrate to Britain from Scandinavia for the winter. For birds the advantage of being in a big group is threefold: protection, food and warmth. The more eyes there are, the more predators can be spotted, and the more food that can be located. And the more bodies there are, the warmer it will be. Flocks are a survival mechanism. They grow especially big in winter, when migrant cousins from the frozen north of the world come south to benefit from our warmer climes.

In days gone by, when grain harvesters were less efficient and farmers didn't plough the stubbles in winter, birds could rely more heavily on the open and empty arable fields, full of the gleanings of the late-summer harvest and abundant with weeds, worms and insects. And this is where you will still find them in winter – great flocks of birds on the open stubbles garnering what seeds, weeds and insects remain. And while most people talk endlessly of the spring as the great time of year to appreciate the birds, there is a large part of me that prefers those enormous and heart-warming winter flocks. It reminds me that we humans are not so very different from the birds at all. Home is the company we keep. And flocking together in hard times is how we survive.

And it was as I stood amid these beeches, admiring the shapes of the trees, and dreaming ancient dreams, that I heard the unexpected song of the wren. Wrens very rarely sing in winter. Maybe it was the warm sun that had set him off? I don't know, but this wren had decided to *pierce the silence with a needle of song*, to borrow some words from the poet Edward Thomas. Wren song is one of the true miracles of everyday nature. And there are very few descriptions that can truly do it justice. Edward Thomas comes pretty close with his 'needle of song'. But the best description I have so far found is, somewhat unexpectedly, from Britain's foreign secretary at the outbreak of the First World War, Sir Edward Grey. (It's odd where you stumble across fellow bird lovers.)

He describes it like this: *The wren's song is a succession of rapid notes, forming a long musical sentence, that is repeated again and again at intervals . . . When a wren is in good form he sings, as it was said the young Queen Victoria danced, 'with decision, and right through to the end'.*

Sir Edward also speaks of a time when he had escaped to a cottage in the country after a 'weary week in London' (nothing changes). He was soaking up the beauty of his surroundings when *a wren sprang into the air, and, singing in ecstasy as he flew, passed straight over me and over the cottage roof to some other place of bliss on the farther side: 'like a blessing' said one who was with me.*

I was overcome by that exact feeling on hearing the wren amid the cold beech trees on that glorious wintry day. It was a blessing.

Sir Edward wrote a book called *The Charm of Birds*, and it became a real inspiration for me during the months of Mum's illness. It amazed me that I could find so many parallels with this man. A man who had lived so many generations before and in another age entirely. I found it incredibly moving and gently reassuring, too.

Sir Edward talks regularly of leaving London to take refuge in his small cottage by the River Avon in Wiltshire, not far from where I live now. An entire five pages can be taken up in his description of a spotted flycatcher's nest or the unusual behaviour of five grey partridges he spots by the side of the road. Or indeed the wren. This was a man who had the weight of a world war on his shoulders but was still able to find time to pause and admire the wildlife around him, to take solace from it and, most importantly, appreciate every delicate detail. He saw beauty *from the eye of a wren to the starry floor of heaven*, as the writer and journalist Ivor Brown once wrote of William Shakespeare.

In fact, I would venture that Sir Edward could not have done his job without this innate appreciation of the natural world on his doorstep. His writing reinforced in me the desire to find out more about the birds around me – to understand the depth of their power to heal, and to learn more about the regard and reverence that our forebears held them in. The wren in particular.

Despite its small stature – and general sweet dumpiness – the wren was regarded by our ancestors as the 'king of birds', which I had always attributed to some kind of weak sense of ancient irony. But not at all: according to Greek legend the wily wren hid under the eagle's wing in order to take the regal crown in the race of all the birds. So upset were all the other birds about this rather intelligent trick that the king of the birds is forced to hide deep in bushes (and my wall) and build dense circular nests with tiny holes for access because, the legend has it, the other birds will kill him if given the chance. And this might also explain why the wren's Latin name is *Troglodytes* or cave dweller. Although I prefer William Wordsworth's interpretation of the wren's nest:

> Among the dwellings framed by birds
> In field or forest with nice care,
> Is none that with the little Wren's
> In snugness may compare.

Wrens seem to crop up in every culture, in every age, as a figure of deep superstition: loved, feared and, in parts of Ireland, 'mortally hated'. I would find it quite hard to mortally hate a wren but then I'm not Irish. Legend has it that a wren beat a drum to give away the location of an Irish army that was subsequently slaughtered by Oliver Cromwell.

Some of the other superstitions are equally bizarre and unfathomable. In some cultures until recently wrens were hunted on St Stephen's Day, 26 December, and then stoned to death. Again, this is because a wren is supposed to have betrayed Jesus Christ to the Romans by singing in the Garden of Gethsemane. It strikes me that this beautiful small bird has been bestowed rather an unfair reputation by some of our ancestors. Though life is never simple; people also believed that it was incredibly bad luck to harm a wren, and all sorts of awful things would happen to those who did, like illness, death or even your hands shrivelling up.

I could quite agree with that sentiment the day I heard the wren's ecstatic song on top of the vale surrounded by those ancient beech trees. I crunched around the leaf fall in search of the author of this angelic sound, looking for his 'other place of bliss', and then I saw his tiny form flitting noiselessly across the branches, seemingly oblivious to the bitter cold. I wondered whether he was one of my wrens from home, living his nights inside my comfortable wren-lined wall? I hoped so. Because no matter how alone or out-in-the-cold you feel at certain points in your life, the knowledge that you have a warm home in which to seek sanctuary means you can withstand pretty much anything. I had always been lucky enough to have that. But for how much longer? My visit home earlier that day had demonstrated the bald reality of my situation: my sanctuary was empty. And I needed to find a new one.

I couldn't shake these thoughts from my head. The sense of impending loss – that cold empty kitchen – and what it would mean for my family. Whenever I had a quiet moment, it nagged away at me. What would the future hold for us? What kind of a future would our little family have without Mum?

And then, a few mornings later, as I was soaking in a Sunday bathtub, I heard an interview on the radio with some high-profile figure or other who was talking about exactly that: the impact of his mother's death on his family. He spoke of how it had fallen apart. All their relationships, he said, were somehow defined and reinforced by this singular North Star. And when the North Star disappeared, each family member just kind of spun out of orbit into the cosmos. I can remember thinking so clearly: *That's us. That's Mum. What will we do when she's gone? But she won't go. She'll keep on.* I could not allow myself the possibility of considering what life would be like without our North Star.

I can look back now, with a wife and children of my own, and truly understand the passion with which Mum built our little world, and the fierceness with which she defended it. And on that cold winter day, as she lay fighting for her life in some unfamiliar bed, I was on the brink of losing it all. I just didn't know what the future held for my family or me. One of the biggest realizations I had, after her admission to the dreaded Ward 4C and my forlorn trip home, was that homes are not made of bricks and mortar. Homes are people.

I'm not sure there is a better tribute to the home my mother created for her family than that when she was absent it ceased to exist. As is often the case, it takes a romantic poet to best explain the situation. In this case Wordsworth in his ode to a wren's nest:

Rest, Mother-bird! and when thy young
Take flight, and thou art free to roam,
When withered is the guardian Flower,
And empty thy late home,

Think how ye prospered, thou and thine,
Amid the unviolated grove
Housed near the growing Primrose-tuft
In foresight, or in love.

Our mother wren created an unviolated grove for all her family to live and thrive within. And there are no words that can express quite how lucky that made us – or how much we feared losing it.

Wren

What it looks like: *The wren is a small dumpy-looking songbird about the size of a golf ball. It has a dappled chestnut appearance all over, with a small stripe above its eye and a stiff, perpendicular tail. People sometimes mistake wrens for mice, because they very often dart out from cover just above ground level.*

What it sounds like: *Its song is a long trilling and varied series of notes, which are very loud and belie its small stature. When it's alarmed the wren will utter a harsh, rattling call.*

Where to find it: *Wrens can be found in any hedgerow, woodland, scrub, garden or park all over Europe – in town and country. Anywhere, in fact, with thick undergrowth. They are one of the most widespread of all songbirds but do suffer heavily in cold winters, when up to 25% of them can die. They construct little globe-like nests in holes in walls or trees.*

What it eats: *Spiders and insects are their most favoured meal, though the wren will eat a variety of grubs and bugs.*

Chance of seeing one: *You have a 90% chance of seeing one in a park or your garden. The wren is one of the most prolific birds in Europe, with tens of millions of nesting pairs.*

4. Song Thrush

That I could think there trembled through
His happy good-night air
Some blessed Hope, whereof he knew
And I was unaware.

Thomas Hardy,
'The Darkling Thrush'

My heart was shaken with tears; and horror
Drifted away . . . O, but Everyone
Was a bird; and the song was wordless; the singing will
never be done.

Siegfried Sassoon, 'Everyone Sang'

There is a song by Paul McCartney called 'Blackbird', which he wrote in 1968. It's rather beautiful. Put it on now, if you can. The opening lyric describes a blackbird that sings in the middle of the night. Now, I don't want to sound like a pedant, Paul, but the bird you're most likely to hear sing in the dead of night is the song thrush and not the blackbird. In fact, not just in the dead of night but in the dead of winter too, when no other bird sings at all.

The greatest joy of the song thrush, and his close cousin the mistle thrush (on which more later), is that unlike most birds, their song rings out at the gloomiest of times. Long before spring's dawn chorus echoes hopefully across the land, this fine-featured and elegant songbird, with its buff-coloured back and creamy freckled breast, will be belting out his cheerful tune. It acts as a powerful restorative to any fortunate soul who is lucky enough to hear it on one of those dank, bone-achingly cold mid-January mornings that do such a good job of sucking out any remaining dregs of optimism or *joie de vivre* we might have had left over from Christmas. The song thrush will pierce the air with a cannonade of sweet song at winter's lowest moment, and just when you'd given up all hope of the spring ever returning at all. I can tell you this with absolute certainty, because I was lucky enough to experience exactly that on a bleak midwinter morning when I was consumed by self-doubt and fear for the future.

People always tell you that 'life goes on' when bad things happen. It's supposed to be reassuring. But it doesn't always feel like that. That relentless sun rises again each morning, and the moon at night. Other people still go to work (how could they?), have parties, get married, buy houses and cars, walk their dogs. The planets keep revolving, selfishly unaware of your plight. And the tide insists on coming in and going out twice a day, as if to rub it in. It's very easy to take W. H. Auden's view when you find yourself at the eye of a personal storm: *The stars are not wanted now: put out every one, Pack up the moon and dismantle the sun.*

It was shortly after Mum had died. I was lying on my bed at goodness knows what hour (I was working hard not to look at my watch) and my mind was whirring at

200 revs per minute – unwelcome thoughts stumbling over themselves to invade my peace. Outside the temperature must have been well below zero, but here I was in a sticky cloak of clinging sweat. January is the month that has no mercy. It sifts out the weak from the strong. It culls. My darling mother had been a victim of this callous month. It was not the cancer that had killed her in the end but pneumonia.

My chief source of anxiety that night was what more we could have done to help her, brought her more comfort at the end. Or how we might have acted differently and maybe extended her life. Could we have found more effective treatment elsewhere? But would we really have wanted to extend her suffering? In fact, should we have taken the decision to end her life sooner? Was it cruel to have kept her alive in such pain? Was it selfish? Could we, should we, would we? What if, what if, what if? It was all so utterly futile this bombardment of guilty introspection. I can see that now. But that's not how you think in the immediate aftermath of a battle lost. To be honest I still wonder about some of the decisions we made, and whether we did the right or wrong thing.

I was idly thumbing through an old notebook I came across recently, as I was clearing out a desk, and I found these scribbled notes. I've absolutely no idea why I took them or why I kept them (possibly some sort of journalistic instinct), but they do offer me some insight as to why we made the decisions we did, or at least one decision in particular. It was just before Christmas and Dad, Richard, Katie and I were squeezed into a tiny consultant's room off Ward 4C – a windowless little box with the heating switched on way too high. Mum's doctor was speaking. At the top of the page I'd written in thick letters: *Home for Christmas Target*. What follows is a kind of checklist of hopelessness.

Lost albumin (marker of body metabolism)
Low neutrophils
Trephine (a hole saw for extracting bone marrow)
Not antibodies, penicillin most likely cause
Restless legs: drugs?
Phenytoin (anti-seizure)
Blood transfusion on Monday
Steroids tomorrow (will improve left side)
Swollen saliva gland (infection or a stone in the duct)
Infusion of white cells . . . can help a little
No chemo ~~ever~~ for the foreseeable future

How much more could one small human take? Not much was the answer. At the end of the meeting the doctor was candid. It was time for a hard truth or two. She knew by this stage that all the children were resigned to the fact that Mum had no hope of survival beyond the next few weeks, but our father just simply would not accept it. His obstinacy, or, I should say, his blind optimism, in the face of all evidence to the contrary was unwavering. And so he had to be told. In no uncertain terms. 'She'll be lucky to survive Christmas, Mr Corbett.' This was the night of 23 December. And that was the day Dad finally accepted, against all his instincts, that Mum was never coming home. He cried.

And I had never before seen my father cry. None of us had. It was an awful moment because I'm not sure what I felt more, as I watched my father, this great ballast tank at the heart of our lives, diminish visibly in body and spirit – sympathy or fear. Deep sympathy for this broken man, my father, hunched over before me, weeping – or a dreadful fear at losing our family's stolid and reliable (if infuriating) figurehead?

'Don't worry, Dad. We'll make damn sure we get her home for Christmas,' one of us said.

That was our goal. One. More. Family. Christmas. All the family were summoned: children, uncles and aunts, various cousins, grandchildren. We were going for the whole nine yards. The hospital was not at all keen to release her. And I don't blame them. She was simply not well enough to go. But Mum was determined. And we were determined. Christmas without our mum was an impossible scenario. In our childhood, Mum *was* Christmas. It simply didn't exist without her. And we knew that this was our last chance. And besides, her dog Quink would never have forgiven us if she could not have had one more day with Mum. Was it the right thing to do? Medically speaking, possibly not, but we were going to kidnap her if anyone tried to prevent us. To that end we hired a private ambulance to ferry her back and forth, invested in a special bed and paid for a private nurse for the day.

We got her back. It was touch and go whether the hospital would release her at all that morning. But they eventually let her go, and Mum was home by late

Christmas morning. She emerged from the ambulance with, in the words of my brother, 'a smile on her face the size of the Forth Bridge'. Dad carried her now tiny bird-like frame upstairs and settled her on the bed. And for the first time in so many long weeks she seemed content. She kept repeating, over and again, 'Home, home. I am home.' Quink, that wisest of spaniels, very delicately laid herself at Mum's feet and did not move for the rest of the day.

A brittle happiness pervaded. Mum was heavily drugged on her arrival so it was hard to communicate at first, but as the drugs wore off more and more of the old Mum began to emerge. 'Like a fleet of boats appearing gradually out of the mist,' Richard said. She received a steady stream of visits throughout the day, and despite our 'one in, one out' policy it was a hell of a lot for her to take. She was strong in spirit but so very weak in body. A little queue formed on the stairs up to her bedroom, each of us waiting our turn to go up and pay our respects. It makes me very sad to think that I cannot remember any particular conversation I had with her that day, any particular wise words or profound insight I could carry with me for years to come. I can't even remember what I bought her for Christmas, which is an awful confession, since Mum was the best buyer of Christmas presents in the land. She only ever bought people what they wanted. And she hated the futility of giving a random present for the sake of giving a present. And yet that day of all days it seemed such a futile act to buy Mum a present, a present we knew would be of no use to her. But I did. We all did. Of course. And those presents sat by her bed in a forlorn little pile.

We did our best to recreate for Mum a Christmas of old. Dad wore the brave mask and ticked as many of the Christmas boxes as he could: the tree in the hall, cards on ribbons hanging from the ceiling, mistletoe in its usual spot – dangling from the lintel of the door to the dining room – a disproportionately large turkey, lovingly cooked by Katie, and all the trimmings that we remembered from childhood.

To the untrained eye everything seemed as it should, but the lights were dimmer somehow, the decorations flagged and the mistletoe hung drably from the lintel, unused. The sparkle was gone. Because the sparkle was lying in a bed upstairs, fighting for breath. The overriding memory I have of that day is of the family downstairs and Mum upstairs. No matter how hard we tried, the unadorned fact of the matter was that the mother we knew, and the Christmas we knew, was absent. And yet despite our flagging morale, particularly at the end, we were lifted by the sure knowledge that we had given our mother a happy day – amid all the gloom – and that she had been surrounded by all the people in the world she loved the most. *I am home. I am home.*

The ambulance came to take her away at 6 p.m. Bright neon and silent flickering lights set against the blackest of winter skies. They wheeled her in. They slammed the doors. And she was gone.

Mum died on a Sunday. And although we knew it was coming, we'd known for weeks, when the moment actually did arrive, as forewarned, in full Technicolor, it still felt like a bolt from the blue. You see, Mum had always had this kind of life force about her – she was one of those infuriatingly vigorous people who never seemed to get ill. Strong, vital and equipped with a smile that would light up Fingal's Cave. We had all just assumed she'd live till 103. But she didn't. She lived till sixty-six instead, and it broke all our hearts.

I arrived at the hospital that Sunday morning with Mary. We'd bought a coffee from downstairs, as we always did, found a newspaper for Dad, who we knew would already be by Mum's side, and were anticipating another long day of holding Mum's hand and giving what comfort we could.

I knew the instant I walked through the door of the ward that she was dead. It was the look of complete panic in the nurses' eyes when they saw us. I rushed headlong towards Mum's little cell, myopic, as the kind nurses melted away. One of them was crying. Dad was hunched over Mum's body, gripping her hand, and staring with startled disbelief at her face. It was only minutes since she had died. And he had been there, by Mum's side, for her last breath on this earth. He looked up, tears in his eyes. 'She's gone.' And then he looked down again.

I don't know how long we were in that little cell next to Mum's body. At some point Richard had joined us. I cannot remember exactly what thoughts I had, except that I understood so entirely then why people talk about the dead as *the departed*. Her body remained, but it was an empty vessel. Her spirit was gone. And there was just no point in staying in that room any longer.

'She's not here any more, Dad,' I think one of us whispered quietly in his ear.

Dad was very calm. He kissed Mum on the forehead and said, 'You were the best wife I could ever have had,' and walked away.

We made our way out to a little anteroom down the corridor, where Mary was waiting for us. I have never been so glad of her reassuring smile and warm embrace in my life.

The rest of the day is a bit hazy for me to recollect with great accuracy. There was quite a bit of hanging about in that airless waiting room, making calls. Katie first. After weeks at Mum's side she had had to return to Edinburgh the day before to take my niece and nephew back to school. I felt awful for her. I knew how much it would have destroyed her not to have been by Mum's side in those final moments. After all those nights and days she had spent at the hospital in the preceding weeks and months, making sure our mother was comfortable and in no doubt of our love. All those thousands and thousands of miles she had driven, trying to juggle two small children and a terminally ill mother. Throughout her adult life rarely a day went by when Katie didn't speak to Mum. She was the railway track guiding Katie's locomotive. They depended on one another: help, advice, gossip and succour. A proper mother-and-daughter act. There was the occasional derailment, of course, when the signals failed, but their souls were fundamentally intertwined. What on earth was she going to do without her? What were any of us going to do?

Later that day we congregated at Richard's house,

about thirty minutes' drive from the hospital. Mary cried in the car all the way. I didn't. I seemed to be experiencing some form of emotional blockage. It would take at least a week to clear.

Richard lit the fire and we stood around in stunned silence, drinking one gin and tonic after the other. Nobody really knew what to say. Strangely, the overriding feeling I had was not one of despair or heartbreak but cold relief. It was over. For Mum. For her family. For me. It seemed like a terrible emotion to feel at the death of someone I loved so much. I always used to detest that dreadful saccharine phrase people used when someone had died of a drawn-out illness: 'It was a relief, though, at the end, wasn't it?' And if someone had said that to me then, I'd have probably told them to bugger off, or worse. But when it came down to it, in truth, it *was* a relief.

We collectively moved on to autopilot and anaesthetized ourselves with gin. Gin and tonics, in fact, are one of my enduring memories of the days and weeks after Mum died. Just so many bloody gin and tonics (and quite a lot of chain-smoking, too). My father didn't stop drinking gin for about six months afterwards. He fell into a kind of impotent despair – a frozen torpor – sustained by gin and bananas. My father, as I remembered him from my childhood – confident, colourful and in control – his machine-gun-like observations on life ratt-tatt-tatt-ing cheerfully across any private or public house – died in that stifling little room next to Ward 4C. I hoped with all my heart that that cheerfully cantankerous, endlessly noisy, infuriating man might return one day.

'I'm not sure what is worse,' I remember Richard saying. 'Mum dying or what's happened to Dad.'

Our family's other foundation stone, the man we knew we could always rely on to get us out of any fix – though not conduct a grown-up conversation beyond 'Where are the dogs?' or 'Have you shut the gate properly?' – had cracked neatly down the middle, hewn in two by the death of the only woman on the planet who truly understood him – or had the patience to put up with him.

Grief's a funny old business. It's one of those hopelessly inadequate words that we use in a feeble bid to describe a host of emotions that tend to manifest themselves in a hundred different ways. Like love, I suppose, but awful. And it takes people in different ways. For me, at first, it affected my dreams – some better than others. The night after Mum died, I had a powerful and achingly vivid dream about a winding leafy lane in autumn on a hill near my childhood home. I could hear but not see my mother. She was just round the corner, up the lane a bit, reassuring me that everything was OK. She seemed happy and content. 'Please don't worry, darling,' she said. The cynic in me dismissed it at the time. Of course these sorts of dreams happen when someone you love dies. But it's never left me. It helped me in the days and weeks afterwards. It helps me still.

Richard, on the other hand, took it one step further. Not satisfied with taking solace from reassuring dreams, he was determined actually to speak to Mum. And with this in mind he visited a spiritualist in a housing estate near Basingstoke.

'I am getting an elderly man,' she told him.

'Erm, I'm here to see my mum actually,' he said.

'Was your mum by any chance a bubbly woman?'

'Oh God,' he said, and left.

My brother has always had a deeply sentimental side to him, and not just when it comes to humans. As a teenager his bedroom resembled more a zoo than a place to sleep; up to and including his Jack Russell terrier, Fly, there was a cockatiel called Baldrick (who was replaced in later life by an African grey parrot called Hector), a chinchilla called Bob – housed in a seven-foot-by-three-foot cage in the corner – and a Russian dwarf hamster named Vladimir, who lived free range and nested in the bottom of the wardrobe. One day Vladimir went missing. This was a monumental crisis in my brother's life. A housewide hamster hunt was instigated that, I am sad to say, yielded nothing. I can remember the floorboards being prised up and friendly mousetraps set up all around the house. 'What if he's been mugged by a gang of field mice?' Richard fretted. And then, when all we had caught in the traps were indeed live field mice and not Russian dwarf hamsters, instead of drowning them in a bucket of water (my father's policy) Richard would take these mice down to the farm and release them. 'Though I'm a bit worried they won't know anybody.'

In the weeks after Mum died, Katie, Richard and I were sent books about grief. Dad was sent lasagnes. His friends clearly understood the practicalities of a man of a certain generation losing the woman in his life who knew where the plates were kept.

'I've lost my wife,' Dad kept saying to me. 'You've got a wife – you're OK. Who is going to look after me?' He was right, of course. From that moment on Dad shot to the top of our collective worry list.

And so it was, a few weeks after that horrible day in Ward 4C, that I lay in my bed chasing thoughts like a furious bloodhound, trying to piece together the last bizarre few months, where my secure little life, full of purpose, meaning and direction – where every step was a step in the right direction – had quite suddenly been sliced, diced and thrown into fate's giant Kenwood Mixer. And then a dear old song thrush struck up. He didn't care about my problems, obviously. He'd most assuredly got his own life-and-death struggles that winter, but he chose that morning to settle in a tree outside my window and start to sing.

There are two ways to take a song thrush singing at the top of its voice outside your window at 4 a.m. on a winter's morning. One is to reach for the nearest heavy object and lob it at said thrush, cursing his barefaced cheek at keeping you awake when you've got *so* much to do today. The other strategy is to wallow in that song. To let it massage your soul. To lie there contentedly and let the notes flow over you. So that is what I chose to do that morning. And not just that morning but every morning of the grim month or so after Mum's death, as the delicate thrush sang its beautiful, sweet melody. Whenever my thoughts overwhelmed me at some ungodly hour in the morning, I knew my song thrush would be back. Living, singing proof that life goes on. Whether we want it to or not. Nature knows. It probably doesn't know it knows. But it knows.

As I write these words, on a cold April morning on day god-knows-what of 2020's coronavirus lockdown, nature has bestowed upon me a mistle thrush nest. It's twenty feet up in the sycamore tree outside my window. I can see it now if I crane my neck. A mistle thrush is a larger, more muscular version of a song thrush – like your more streetwise cousin who's into contact sports. It's called the mistle thrush because it's extremely fond of the berries on mistletoe bushes. Simples. I've been absolutely transfixed by these two mistle thrushes building their beautiful raggedy nest. It's a work of art. And watching its construction by those two busy, characterful birds completely takes me away from myself. The mistle thrush, like its song thrush cousin, can also knock out a dazzling tune. It's famous for singing its way through even the blackest of winter mornings. Our wise ancestors commonly referred to it as the 'storm cock' because it is the only bird that keeps singing no matter what the weather throws at it. A useful metaphor for us all, I always think.

People have banged on about the beauty of their song for centuries, but just how do you separate the sound of the song thrush, blackbird or mistle thrush from the cacophony of other birdsong ringing around the valleys and towns? Truth be told, it's bloody hard to tell them apart at first. And it'll take the uninitiated ear a while to work it out. In fact, when I started out I found it incredibly hard to differentiate between any songbird's songs.

Looked at in the round, if there were an *X Factor* for

thrushes, then I suspect the blackbird would win the cash prize and dodgy record contract. The blackbird's song is undoubtedly more lyrical than that of the song thrush, and when you hear its first hearty notes in late February it feels like a bowl of warm honey-infused porridge on a cold day. It also behaves a bit like a spoiled twenty-something pop idol. To look at, beautiful, with its inky black sheen and bright yellow beak (the female has a more sturdy, hardworking brown colour to it), but it probably wouldn't be very good company if you took it out for a drink: bossy, at times hectoring, and fixing you with its steely gaze if ever you said something out of turn. The blackbird, in short, is one of nature's great characters: charging about domestic gardens and parks, in town and country, with a grumpy, proprietorial air. I sometimes feel when I come across one in the lane or in my garden that I've committed some great breach of blackbird etiquette by having the temerity to step into his realm uninvited. He'll squawk furiously from a nearby hedge or tree. But even his warlike, impetuous squawk, which is designed to send me packing with a flea in my ear, fills me with delight. For some reason it always trans-ports me back to cool autumn evenings, long shadows and the smell of woodsmoke lingering in the air.

Like all songbirds, blackbirds have also played a cen-tral role in the human story – in our culture and folklore, though not always in a way the blackbirds would have entirely appreciated. Up until about the sixteenth cen-tury people ate blackbirds (four and twenty blackbirds baked in a pie, anyone?). In fact, the medieval nobility were partial to eating any kind of songbird. They were a delicacy. Later on the Victorians used to capture songbirds and put them in cages to set off their elegant parlours.

Mankind has always been in thrall to the sound of songbirds, going back to the dawn of time. It makes total sense. Before we had record players, CDs or MP3s, it really was the only music most people would hear each day (only the toffs could afford to watch a Handel opera or a symphony by Mozart). And because the thrushes are the loudest and most persistently beautiful of all the songbirds it is no surprise that so much poetry and prose has been dedicated to their ability to rouse our spirits.

When you hear a blackbird sing in the hedgerow or a song thrush in a suburban garden, you can draw solace from the fact that it is a sound that has lifted the spirits of people, standing on that very spot where you are right now, for many thousands of years. Thinking of that always takes me back to a poem by Edward Thomas. For me it gets to the essence of the kind of peace that nature can provide. It concerns a snatch of ethereal calm amid the rush of the industrial age. The author is on a train that draws up at a remote rural station called Adlestrop in late June. *The steam hissed. Someone cleared his throat. No one left and no one came.*

> And for that minute a blackbird sang
> Close by, and round him, mistier,
> Farther and farther, all the birds
> Of Oxfordshire and Gloucestershire.

What makes it particularly poignant was that soon after he wrote those lines, Edward Thomas was killed in the mechanized human slaughterhouse that was the First World War.

I am lucky enough to be surrounded by blackbirds where I live. In fact, right now, as I write this line, I reckon there are at least three blackbirds nesting in and around my own and my neighbours' gardens. Every day as I look out of the window another freshly fledged bird will be learning the ropes – keenly watched by a parent – at the foot of a hedgerow or hopping about a grassy verge. I could watch them all day. And listen to them too, of course.

The blackbird's song is a fruity, fluting sound that massages the ears – not unlike Paul McCartney's back catalogue really. The song thrush on the other hand has more of the Rolling Stones about his song. It's of the same genre – fruity and melodic – but it's a sharper, more persistent, energetic music. Gin-clear. According to the poet Gerard Manley Hopkins, *It strikes like lightnings to hear him sing*. But the biggest difference between the songs of these stalwarts of the rural and suburban garden is that the song thrush will repeat the same refrain again and again. In the words of Robert Browning, in his poem 'Home-Thoughts, From Abroad':

> That's the wise thrush; he sings each song twice over,
> Lest you should think he never could recapture
> The first fine careless rapture!

The best way to learn birdsong – and differentiate the song thrush from his blackbird and mistle thrush cousins – is to go out and about in early spring, when the birds are singing and there are no leaves on the trees. Once you see a blackbird, song thrush or mistle thrush in action, you won't forget it in a hurry. And there are a million and one apps on your smartphone that can help, too. Although I've made that sound a lot easier than it is. I remember setting out on my mission to learn the authors of all the birdsong around me and giving up almost immediately as I was quickly overwhelmed by the sheer quantity and diversity of the noises I came across. But I persevered, and it was worth it.

Learning birdsong is a bit like starting a book by Hilary Mantel. The initial prospect of ploughing your way through 1,000 pages feels like an impossible task, a mountain too high to scale. *My kids will have grown up and I'll be in a retirement home by the time I finish this*, you think. But as soon as you get past the first chapter, suddenly it becomes a joy. You can't put it down. You skip through the pages with happy abandon and long for it to go on and on and, in the case of *The Mirror and the Light*, on. And so it is with birdsong. Once you've got your head around, say, three or four common garden birds, then you have a solid foundation for further enlightenment. One day you'll hear a cheerful cascade of song in the garden on a May morning, and you'll think to yourself: *Well, it's not a chaffinch, or a robin – I know those now – or indeed a blackbird or a thrush, so maybe it's some kind of warbler that's just arrived from Africa. I'll put that into Google and see what comes up.* And so the journey goes on. Discovery after discovery. Layer upon layer. Until one day, going outside into the garden – or for a long walk – moves from being an act of basic utility (must get some fresh air and exercise) into an act of unadulterated pleasure. It's no longer you alone in the garden or pounding down the lane; you are just one other creature in a broad and diverse universe oozing life at every turn. You become grounded. Perspective dawns. No longer playing the leading role in your own tragic private melodrama, but merely a bit part in nature's great epic.

The problem, of course, with all this joy-at-life-for-life's-sake stuff is that it means walks and other outside sojourns become rather extended affairs. You lose yourself metaphorically speaking and, on more than one occasion for me, actually. Some philistine once suggested to me that it was all rather pointless learning the names of the birds and what they sound like. 'Why do you need to know? Why not just listen and watch instead?' On hearing this, and having settled the emotions that were raging inside me, I explained it to him in this way: imagine if you lived on a city street or in a village or town and you didn't bother to get to know your neighbours? Imagine how much poorer your life would be for it? Getting to know your neighbours – for good or ill – adds ballast to your emotional life. It adds depth and richness.

I used to visit a coffee shop near my work in London every day for months on end, served by the same cheerful three hipsters. I'm ashamed to say that for at least three weeks I took no interest in them. I didn't learn their names or ask how their day was. I was always cheerful and polite and so were they. And that was just fine. Then one day I got talking to Jonny about his commute and we realized we'd been drinking in the same pub after work. Jonny introduced me to Anna, the boss, who lived in Petersfield, not far from where I grew up. And then there was Timea, who had a bit of a crush on a colleague of mine. Suddenly this whole world opened up. My morning coffee, which had at one time been functional and polite, a precursor to another suffocating day in the office, became an event I looked forward to each day. Another sweet layer of jam added to life's roly-poly. My life became that little bit richer for it. I needed that. And all I'd done to open that door was to learn their names. It's the same with songbirds. It's the same with all nature. Once you introduce yourself to it, you'll never look back. I met my song thrush on that horrible morning back in January 2012, and it made all the difference. I look back now and I cannot even imagine how I got through those darkest of winter days and nights without the song of the thrush to get me through.

It brings to mind that last Christmas with Mum. While incredibly painful at times and the palest of imitations – it was for me, for all of us, a shining beacon of light in the gloom. It was a vigorous song thrush singing on the darkest of nights. Together, my family and I achieved something that had felt totally impossible, and I can for evermore look back upon that day, my mother's last Christmas, and hear her words echoing softly in my head, repeated over and again like a song thrush's refrain: *Home, home. I am home.*

Song Thrush

What it looks like: *A fine-featured medium-sized bird with a buff-brown back and mottled creamy breast with black spots. Mistle thrushes are very similar but are larger, more muscular birds.*

What it sounds like: *A loud and melodic fluting song – quite persistent – that can often be confused with a blackbird. The way to tell them apart is the song thrush will repeat the same phrase again and again. If it were talking to a snail, it might say, 'I'll eat-ya! I'll eat-ya! I'll eat-ya!' They are known for singing all through the winter from December on, especially noticeable on dark mornings.*

Where to find it: *Song thrushes nest in trees and bushes, very often around houses and gardens, which is where you are most likely to see them. They move about on the ground with a characterful little hop-hop, and can be seen digging for worms or smashing snail shells against stones.*

What it eats: *Earthworms and snails are favourites. It also eats insect larvae and berries.*

Chance of seeing one: *It used to be that most rural and suburban gardens had a song thrush nest in it, but the population has crashed over the last forty years for reasons that are unclear:*

overpredation by cats, corvids and sparrowhawks, and a lack of food thanks to pesticides killing grubs and insects are the most likely causes of their decline. If you are in a lightly built-up area — the suburbs or small towns and villages — there is about a 50% chance of hearing a song thrush in late winter, spring and early summer.

5. Bullfinch

'Beyond the Wild Wood comes the Wide World,' said the Rat. 'And that's something that doesn't matter, either to you or to me. I've never been there, and I'm never going, nor you either, if you've got any sense at all.'
Kenneth Grahame, *The Wind in the Willows*

'I am angry nearly every day of my life, Jo; but I have learned not to show it; and I still hope to learn not to feel it, though it may take me another forty years to do so.'
Louisa May Alcott, *Little Women*

If my father were a bird, I think he might be a bullfinch. Not because he particularly resembles the bullfinch, with its 'bull head', black cap, stubby beak and bright blood-red breast, but because they share similar habits (admittedly they're both quite stout). Bullfinches tend to hide themselves away from mainstream bird society, often at the heart of thick hedgerows, and, despite being one of the most widespread of all the world's songbirds, the British versions are known for never really leaving their immediate surroundings. And so it is with Dad.

My father has never owned a passport, never been abroad, and lives today, aged seventy-five, not twenty miles from where he was born. Actually it's not quite true to say he's never been abroad: he went for a day to Holland in 1983 on a temporary passport. I forget why. Mum probably forced him to go at gunpoint. On his return he handed back this temporary passport to the border official and to the official's mild bewilderment said, 'Keep it. I won't be needing that again!' And he was true to his word. But despite his desire never to fledge further than the parishes and boundaries of his childhood, he is extremely content. He is an expert in his own environment – totally in harmony with it – and knows it better than any other.

I once tackled Dad head on about why he never went overseas. It used to frustrate me. 'How can you possibly *not* want to see the world?' I said. 'I've got plenty of world where I am, thank you,' he said. 'And besides, if I go abroad I won't be able to take my dog.'

Most decisions in my father's life revolve around the comfort and happiness of his dogs and his sheep. As I say, this used to frustrate me, but now I am beginning to think that, just maybe, he had a point.

Why this constant urge to rush about the globe ticking off places on the atlas like some kind of deranged geographical conquistador? My question is: what are we missing in life by *not* standing still? If life is like a country lane, surely it's better to walk down it, taking the time to soak up the beauty as we go, rather than furiously pedal or, worse, drive down the lane and as a consequence miss all the glorious life in the hedgerows. What is the point in getting to the end of the journey, proud of all the vast ground you've covered, but not actually having given yourself the time to see and to appreciate – to learn about – what surrounds you? Why does society slightly sneer at those people who prefer to stay at home?

I spent a lot of time at home in the weeks and months after Mum died; I found great comfort in familiar surroundings. One of my most frequent visitors, from about February onwards, was a pair of bullfinches. I had a small cherry tree in my garden, and a rose or two, and these two industrious little birds would feast on any buds that dared show their face. Bullfinches are known for being vociferous bud eaters, and for a long time were regarded as the enemy of gardeners and orchard owners alike. If you read old books about birds, then you'll see epithets like 'mischievous' applied to the bullfinch, and lines like this from a bird guide published in 1861: . . . *the "olph" [bullfinch] commits sad apparent havoc on the blossom buds*. The book goes on to explain the various ransoms offered for the head of an olph 'paid out of parish funds'.

Olph was an old name for the bullfinch. In other places it was called a 'budding bird' for obvious reasons, or a blood-olph on account of its blood-red breast, which I much prefer. I have absolutely no idea of the origin of the word 'olph'. It sounds like the noise you might make when you bump into a stranger with a hot cup of coffee in your hand. Though one theory I dug up, and quite liked, is that 'olph' is an ancient word for 'elf'. Blood-elf has a good ring to it.

I became utterly obsessed by these bullfinches. I talked to people of little else. I formed a limited company (part of a plan to become master of my own destiny at some point in the future) and called it Bullfinch Media. I stalked those bullfinches whenever I was out walking, listening intently for their gentle piping call emanating from deep within the hedgerows. I tried to photograph them in vain, which is why I am the proud possessor of about fifty photographs showing grey-pink-red blurs in the far distance. It also became such a 'thing' for me that I am now forever associated with the bullfinch by the people close to me. If Mary sees anything bullfinch-related (postcard, painting, key ring, book) she will immediately buy it for me. I still can't quite fathom why I developed such a love for these cheerful stocky little birds. Maybe because they are so beautifully intent: practical, busy and thorough. Or is it that they are somehow unreachable in their thick hedgerows?

Another feature of the bullfinch character is that males and females form long-lasting attachments, and whenever you see a single bullfinch you are sure to see their mate a few feet a way. And so it was with my father and mother. In all my childhood I don't think they ever really spent more than a week apart. They were inseparable. Devoted to one another. Nobody could ever quite see what my mother – the beautiful, intelligent and well-travelled reader of books – saw in my father – old-fashioned, opinionated, culturally blinkered, and who'd never read a book (and still hasn't). Even Dad's own father simply could not understand what my mother saw in him. And he wouldn't have won too many awards in the 'good husband' category either – certainly not in the early days.

In fact, if there were a ranking for the most unreconstructed 1970s husband in the world, I think he'd finish a short second to Andy Capp. As far as I can remember he spent most of my childhood keeping as far away from his young children as humanly possible, very often in the varied pubs littered across the valley where we lived or undertaking sporting endeavours with his friends.

He was not in any way deliberately unkind. Merely absent. He just lived his own life. The only time he ever got really furious was if he caught one of us in his dressing room, rifling through his desk in the sitting room or borrowing something without permission. 'You're a bloody communist!' he'd shout if he caught my brother or me borrowing a pair of his socks. 'Buy your own damn socks.'

It was also a great relief not to have one of those fathers who put pressure on his children to succeed or tried to mould them into mini versions of themselves. Quite the opposite: Dad's life and his possessions were very much his own. They were *his* toys, so go and play with your own toys. This was the case with his real-life farm, as well as the 1950s toy farm he kept from his childhood and which fascinated me as a boy. It was hidden away in a deep cupboard, far from prying eyes, and it was a very rare treat that he would allow me to play with it – under strict supervision, of course. Even today if I ask him a question about the running of the real farm – the price he was paid for his lambs this year, for example – he will cross his arms defensively and say, 'That's none of your business!'

'But, Dad, for crying out loud, surely you want one of us to run the farm one day? Don't we need to know this sort of thing? I'm forty years old!'

'Trunky want a bun?'

'Oh, for God's sake, Dad, now you sound like Mum.'

When he wasn't prowling around the house, checking his children weren't stealing his toys, or out and about 'somewhere on the farm', he'd be at our tiny village pub – the Hurdler's Arms – more an ancient parlour with a fireplace really than a pub. This pub, where my father spent so much of his time, and where the landlord, Jack, installed a set of swings exclusively for the use of Dad's children to keep them occupied while he was inside, was described by William Cobbett, the great eighteenth-century radical MP in his book *Rural Rides* as 'not half so good as the place in which my fowls roost'. But it was good enough for my father. And Jack didn't allow children inside, which was even better.

I can remember vividly the tragic day in 1993 that the Hurdler's Arms closed its doors for the final time. Dad went into a kind of depression for weeks on end, as did Jack's other handful of loyal regulars. They were like tortoises without their shells. I saw Mum a few weeks after, expecting her to be elated at this development, and yet she too seemed strangely upset by it all.

'But, Mum, surely you must be pleased the pub has shut? Dad'll have to stay at home now,' I said.

She looked at me gravely. 'Not likely. At least when the Hurdler's was open I always knew where he was.'

And yet despite all the ups and downs and Dad's

questionable domestic skills there was genuine affection between them. They were one of those symbiotic couples that were only spoken of as a single entity. And while the rows were sometimes big, so was the love.

I have fond childhood memories of my father, but I certainly could not admit to being particularly close to him in emotional terms. I'm not sure if I've ever shaken his hand, and I've definitely never hugged him or had one of those deep and meaningful father–son conversations you see in American movies. Simply put, he's of a generation, and a family, that didn't do that sort of thing. It never bothered me at all growing up. He was a reassuring presence around the place, trundling about turning off lights that had just been switched on ('Do you need it on?'), unplugging electrical devices ('Fire hazards!'), and thwarting my mother's attempts to turn up the central heating in winter ('Put another jumper on!'). When it came to domestic comfort, his idea of the height of luxury amounted to hot and cold running water. As Mum used to say, 'The problem with your father is he's all barbed wire and baler twine.' And on more than one occasion I seem to remember her yelling at Dad, 'It's not the bloody nineteen fifties, Peter. People just aren't prepared to live in shitholes any more!'

But he would always be there to pick up the pieces when things went wrong. He never judged us when we got into trouble at school or castigated us when our terrible reports flopped on to the mat with alarming regularity. (He himself had hated school so he was most definitely onside when it came to difficult teachers.) For example, he didn't give a damn when I was suspended from my boarding school for being caught drinking premium-strength cider (bought with a fake ID) and smoking cigarettes in a disused railway tunnel some miles from school; he was even a little proud, I think.

'Do you know what the last words my housemaster at school ever said to me were?'

'No, Dad, I cannot possibly imagine.'

'Let me out.'

'And do you know what my last words to him were?'

'No, Dad.'

'Nothing. I just walked away.'

So Dad had no qualms about my somewhat lesser offences, but what did irritate the hell out of him was having to take time out of his day to pick me up and drive me home. 'I pay to keep you there!' he bellowed (the collective noun for a group of bullfinches is a 'bellowing'). And then spent the rest of the drive home telling me what a terrible school it was and how he was going to write to the headmaster and get a refund for the week I was to spend at home (the selection criteria for sending my brother and me to that particular school had been that it was a 'pretty drive' to drop us off and pick us up every six weeks or so).

He also took a dim view of further education. When I announced I was to study history at university (and not become a farmer) he said, 'What are you going to be, a bloody librarian?'

The thing I've learned to understand about my father is that he is a deeply practical man; I am deeply impractical.

'Why would you not, if you were intent on wasting four years at a university, study something that could be of some use?' And while I don't necessarily agree with the sentiment, I can completely see his point of view.

Like the bullfinch, up until the year or so after Mum died my father conducted his life with a kind of busy intent. He never dwelled. I've always envied that in him. He moved busily through his life without pause or inward reflection. In Dad's world there was always another job to be done somewhere. He worked incredibly hard and as well as running his 200-acre farm he was a successful auctioneer and land agent. When finally at night he'd kissed his sheep goodnight, checked on the horses, shut the chickens up and switched off every light in North Hampshire, and unplugged everything else, he would collapse in his chair with his dogs at his feet – and sometimes our African grey parrot on his tummy – and sleep. He lived entirely in the moment.

When Mum was ill in hospital, this practical day-to-day approach to living served us and Mum very well. You would not have found a more loyal and supportive bullfinch. Each day he would trundle up to Ward 4C – all tattered tweed and Terramycin stains – with armfuls of newspapers and *Farmers Weekly*s, sit down next to Mum's bed and settle in for the long haul. I'm afraid the poor nurses could probably smell him coming. Without Mum in the house to put him straight, baths went out of the window ('Only dirty people wash.') and sometimes shirts could be worn for a week at a time ('Why waste washing powder?').

'I'm afraid Dad's gone a bit gamey,' Richard would warn friends before they saw him.

He was sustained at that time by the entrenched belief that life would, of course, return back to normal one day soon. He need only wait it through; grit his teeth and hold his course through this cancerous tempest. His darling wife would be home in a few months, oozing her usual energetic good health, cooking cottage pies for dinner, laughter ringing around the house, and giving bollockings for staying too long in the pub. Nothing that anyone said could convince him otherwise – until that December night just before Christmas, when the cold reality that Mum was never coming home hit him squarely between the eyes. All his busy intent and living in the moment disappeared after that, and he became very difficult.

It's easy to look back after you lose someone you love and create a new story. Who needs the hard truth when you can instead recreate a reassuring soft-focus fairy-tale past and languish in that? I've certainly been prone to this rose-tinted approach; as have we all, in particular my father. We used to give him quite a bit of good-humoured stick after Mum died for his constant allusions to this perfect (nearly) forty-four-year marriage he seemed to be part of. It wasn't one his children recognized. If ever he disagreed with any of us, on anything, he would trot out the same hackneyed line, no matter how mundane the topic: 'Your mother would have been very unhappy about that!'

'Hang on, Dad, you mean, she'd have been unhappy about you having to go to Uncle Ricky's birthday in

Wagamama on Winchester High Street on Wednesday night, when you wanted to stay at home and watch repeats of *Only Fools and Horses* on UK Gold instead? Are you sure that it's not just *you* that's unhappy about that?'

He remodelled himself as an ideal husband: attentive and dedicated to Mum's daily happiness when she was alive. It had been a blissful marriage where he rarely put a foot wrong and he simply could not remember any kind of falling-out and rarely a cross word. 'Your mother and I never argued . . . not properly,' he'd tell us with a straight face.

'So you were never locked out, Dad? And I was dreaming when I had to let you in through the larder window in the middle of the night, when I was six years old?

'We didn't have to hide Mum's car keys under my mattress when she tried to leave you, yet again, after card night at the Hurdler's Arms?

'In fact, the Hurdler's Arms was clearly a mirage and Clayton Ponting had never, ever offered you just one more quick whisky for the road because you've missed dinner now anyway?'

I began to grow a little weary of all this remodelling that was going on of Dad's former life. On one occasion, when I mentioned that Mary was off on a work trip abroad for a week, he looked at me like I'd announced she was conducting an illicit affair with a Spanish salsa teacher. 'Abroad? Abroad? Who's going to look after you? Your mother and I *never* spent a night apart, not in (nearly) forty-four years! Not a single night.'

'What about the holiday she went on in South Africa with cousin James in 1988?'

'Oh, well, OK, there was that week.'

'And what about when Mum took us all skiing in 1984 and you refused to go on account of not liking the French?'

'Yes, well, that was only five days, and your apartment got burgled, so I was right about the French.'

'And Uncle Andrew's wedding in America?'

And so we went on, back and forth in this manner, for months on end. It was never vituperative this back and forth. We laughed about it. As would have Mum. Funnily enough all that Dad's subtle remodelling of his past did was to cement in our minds how much he loved her and how much he would miss her. And the reason I keep writing 'nearly' forty-four years when I mention their marriage is because when it came to the inscription on Mum's gravestone, our dear father told us there just simply wasn't enough room to put long messages about 'loving mother and grandmother' to all of us – and that he would think of something short and meaningful instead. And so the inscription on her gravestone reads: *Margaret Corbett, darling wife of Peter Corbett for nearly forty-four years.*

Despite this got-to-get-on attitude to life and the somewhat belligerent outer shell, at heart I know my father to be a big softy. You only have to watch him with his dogs to understand that. Completely unable to hold a grown-up conversation with his children, Dad can speak for hours to a dog. 'All that love's got to go somewhere,' my brother-in-law, Ed, says.

Dad's dogs have always had official names and then the names that he calls them when he thinks no one is looking. So, for example, his dog Islay was known to Dad as Buzzle Wuzzle, and Lussa, who replaced Islay, was known as Baby Buzzle Wuzzle, or Baby Buzz for short. And when Baby Buzz had puppies they were known as the Wuppers. On his dressing table there are only two pictures: one is of my mother and the other is of his first dog, a small collie called Fuzz. Almost seventy years later he still talks regularly about Fuzz. This is a man who was clearly born with a strong hugging gene but has never felt able to make use of it with other humans.

There was a moment just before Mum's funeral when he was looking rather sad and alone – like a lost toddler in the supermarket of life – and my brother, sister and I had this debate over whether one of us should a) put an arm round him and b) if we did, which of us it was to be. My sister took on the responsibility and gave him a huge life-affirming hug. He stood motionless as Katie draped herself over him, his arms fixed by his side, staring blankly ahead. Richard and I crept up and gently squeezed a shoulder each and then hastily retreated. It was all just very awkward. Mum's death had been like an unexpected car crash in Dad's busily intent life. He hadn't seen it coming and he wasn't wearing a seat belt. His brain could not comprehend a tragedy in his life on this scale.

And so in those weeks and months after Mum died he had no emotional outlet. No framework that could allow him to cope. He was completely at sea. Both he and my mother were brought up at the Bottle It Up and Push It Down School of Taking Hard Knocks. And what a shitty education that is.

But at the time I simply had no tolerance for his grief. I could not accept the form it took. He leaned heavily on all of us. Too heavily, I felt at the time. I can now look back and see that he was not himself, and quite understandably so, but I found it hard to be quite so forgiving. To my mind he had become myopic: unable to see anything but his own misfortune. The world might as well shut up shop right now and set a course for the nearest black hole. This tragedy was Dad's tragedy and Dad's alone. And to be quite frank, to me, after a childhood where he had been largely absent ('Do just try to speak to them occasionally, Peter, rather than just give orders.') it felt like he was cashing emotional cheques he had no right to cash.

I took every opportunity to run away from it all and sat around being cross and angry at the world, simmering quietly, retreating from Dad as he retreated from his old assured self. And while it had never before really bothered me that Dad took no interest in my personal and educational development as a child – and, let's face it, his complete lack of interest at bad behaviour or weak exam results was pretty damn convenient – at the time of Mum's death I became quite furious about it all. Katie, Richard and I all say the same thing when we speak of Dad and our childhood: 'We brought ourselves up.'

As Dad retreated deeper into his metaphorical

hedgerow after Mum's death, the lonely bullfinch, he became increasingly belligerent. He'd call and call and call . . . and call . . . his children – no matter what the hour of day or night. I would look up from my desk and see fifteen missed calls on my phone and at least five answerphone messages with Dad getting increasingly furious on each one, demanding I pick up the phone. When finally I would pluck up the courage to call him back, he would shout, 'Where on earth have *you* been?'

'Dad, I spoke to you last night, and it's now eleven a.m. I've been at home, obviously.'

'Then why don't you ever pick up your phone?'

'Because I was working on something, Dad. But I've called you now. What do you want?'

'I'm trying to track down your sister. She's not picking up her phone.'

'She's in the car, Dad, driving down from Scotland to cook you dinner tonight.'

'Well, as long as she doesn't cook me another bloody lasagne. Your mother knew I hated lasagne.'

And so these somewhat infuriating exchanges would go on. If it was bad enough when Mum was ill, it got worse in the months after she died. We had another invalid on our hands. Except this one was in a very bad mood.

One night Dad had invited some of his old friends round for dinner – I think it might have been his birthday – and the children were summoned. We all rallied round: Richard had no doubt bought some nice wine and Katie had driven the 500 miles from Edinburgh to be there for that night. Katie, a trained chef, cooked roast lamb for Dad and his friends. It hadn't been easy because Dad had taken the opportunity, now that Mum was not around to challenge the decision, to save money by turning down the old oil-fired range we used to cook our meals on. Except he hadn't told my sister. It was only after about half an hour that we realized what Dad had done by which time the joint was in the oven and there was little to be achieved except to turn the range up again and wait. Needless to say, it was all taking a bit longer to cook and Dad was getting increasingly testy, barking at us from his armchair, 'Are we eating this side of Christmas?' And whenever his wine glass got too low, to me or my brother he would say, 'This is a very dry argument!'

Eventually the roast lamb was brought out for the nabob to carve. It looked delicious. Restaurant standard. Nice and crispy on the outside and just the right level of pinkness on the inside. Though Dad had a different view. 'This lamb is bloody raw!' he bellowed. 'Your mother would never have cooked me raw lamb.'

We often have conversations, Richard, Katie and I, about our constant desire to please Dad, and our constant failure. 'The thing you have to remember about our father,' I tell them, 'is that he is a potato. You cannot keep biting into the same potato, again and again, and expect it to taste like a strawberry. We must accept that he will always be a potato, a potato we love, but a potato nonetheless, and he'll never be the strawberry we want him to be.'

In Dad's defence he is at heart a kind man, generous

and scrupulously fair, with that oh-so repressed hugging gene tucked deep down; he has an old-fashioned decency about him. If anyone ever hit difficult times and fell behind on the rent of one of his farm buildings, say, then he would quietly forgo the money and carry on. One of his tenants paid just one year's rent over the course of fifteen years. Dad charged according to means.

Mum's death was pretty much his lowest moment. He simply had no idea how to cope. He is a product of his own old-fashioned upbringing, and in my book to understand all is to forgive all. 'You'll never change him,' Mum used to say to us. 'I know because I've spent forty years trying.' I've learned the hard way that you can never change people, not really, but I suppose you can try to understand them, for your own sanity if nothing else.

Funnily enough, despite his complete failure to cope in the aftermath of Mum's death, my father has always shown the way when it comes to finding happiness in nature. For me it is the birds, for my father it is his sheep and dogs. Never is a man more content than my father when walking around a field of his own sheep dressed in his tatty old blue boiler suit, stick in hand, and wearing his antique cap imbued with the grease of a thousand unwashed heads. And in the months after Mum died I know that those small southern meadows full of sheep, with his dogs by his side, were a deep source of solace for my father. As was Mum's dog, Quink, who went into mourning when Mum had gone but took on the responsibility of looking after Dad with gusto. She adopted him and became glued to his side. They mourned as one, Quink and Dad. In the words of Sieg-fried Sassoon:

> Who's this—alone with stone and sky?
> It's only my old dog and I—
> It's only him; it's only me;
> Alone with stone and grass and tree.
>
> What share we most—we two together?
> Smells, and awareness of the weather.
> What is it makes us more than dust?
> My trust in him; in me his trust.

Those sheep and his dogs were, I believe, for a time the only creatures that truly understood Dad's pain.

I'm reminded, whenever I see my father amid a field of his sheep, of a scene at my wedding to Mary. Mary's uncle – a sculptor – came over to us afterwards, concluding what looked to be quite a long conversation with my father. I was surprised, and not a little concerned. *Oh Lord*, I thought to myself, *what on earth has my father – the ruddy-nosed farmer who wouldn't know a Rodin from a roadwork – been saying to this metropolitan member of the artistic elite?*

'Hello, Antony,' I said with trepidation. 'How's my father?'

'Great!' he replied. 'He's a fascinating man.'

'Really? Erm. In what way fascinating exactly?'

'He has this really *special* affinity with his sheep.'

I'm not sure what I felt more, embarrassment that Dad will have spoken for twenty minutes to this erudite and gifted sculptor about his sheep, and without stopping to ask him a single question about his own life, or relief that this erudite and gifted sculptor clearly didn't seem to mind at all.

I told Mary about it later and she said, 'Well, that'll make a change for Antony.'

Mary has always seen the best in my father. She sees through all that external belligerence and bellowing – in a way I never do – right through to the sweet-natured, quite vulnerable man deep within, cuddling up to his dogs and sheep for security. In fact, so much so that it quite infuriates me at times.

'I just can't believe the way Dad spoke to me just now,' I'll bellow at her, having just put down the telephone to him. 'He just shouts at me all the time.'

'Well, why don't you try not to shout at him or me?' she'll respond. 'I heard the whole conversation and you were just as rude to him as he was to you.'

'Yes, but . . .'

But before I get too angry, I'll reflect that Mary also sees the best in me, too. And, more worryingly, she sees in me what I dare not recognize in myself: that I am my father's son.

One morning in early spring, during the period when Dad was leaning on us the hardest, and I was a seething ball of compressed (and no so compressed) anger, I had my mind altered by a chaffinch.

I was lying in bed when I heard this incessant tap-tap-tapping on the windowpane downstairs. On and on it went. I got dressed and went down, looked into the room where I had thought the noise was coming from, and saw nothing. A mystery. Went into the kitchen. Made a cup of tea. And there it was again. Tap-tap-tap. Rushed back to the room. Still nothing. Days went by. This incessant mystery tapping went on. And still I could find no answer. Mary began to grow incredibly weary of it. Truth be told, it was driving us both a bit mad.

Eventually I decided to ambush the enigmatic tapper. I hid behind a chair early one morning and waited. And waited. After what felt like about three days (probably five minutes) a beautiful cock chaffinch appeared at the window, resplendent in his bright spring plumage – steel-grey cap and copper breast shimmering in the morning light – and shining black eyes. As I was staring at him in wonderment he suddenly began to attack my window-pane with gusto. Tap-tap-tap-tap-tap. And then he'd fly off, take a run-up, and throw himself bodily into the window – again and again. It was an extraordinary dis-play. This chaffinch, it turned out, was vigorously fighting his own reflection. He'd fly past every day – see this wicked intruder on his patch, reflected in my window – and launch a series of exhausting attacks. The futility of it all was incredibly sad to witness. I felt so very sorry for this somewhat dim-witted chaffinch, hammering him-self against his own reflection, when he could have been using the energy to find a mate and build a nest, and do something constructive with his life. And while he was busy fighting his own reflection, no doubt another chaf-finch was stealing his territory and making off with his mate. Sometimes wildlife can give you such enormous clues to what is going wrong in your own life that it is painful. It is an alarming realization to have spent most of your entire life doing your best not to turn into your father, and then for it to dawn on you that he is your exact reflection. I am that chaffinch, fighting my father's reflection in the windowpane. And it took me half a lifetime to work it out.

Chaffinches are one of the first birds to wake us out of our winter slumber. Their song is not hugely attractive, some describe it as grating, but it is addictively cheerful and such a fundamental part of the background of summer that hearing it for the first time in early spring shoots a shaft of untainted optimism to the soul. Hearing the song of a chaffinch in February, and the tingle of optimism it brings, gives me a feeling I could never preempt; it just happens without me trying. And that is what I find so wonderful about nature, and in particular birdsong. You don't have to make any kind of effort to change your mood. Nature just sort of does it for you.

Despite how it might seem the complete opposite at times, in our rushed and stressful lives where we don't have time to notice a meteorite crashing to earth, let alone an early primrose or a bird's song – we remain, as a species, deeply entangled with the natural world in which we are a part. It is engrained in us, whether we like it or not. Little switches go off in reaction to our surroundings day in day out, and we don't even notice it half the time.

I can remember first moving to London in my twenties and working in the heart of the City, near Pudding Lane. It was a late-March day, miserable and cold, with drizzle that seeps into the soul, and cars and lorries thundering down Lower Thames Street, spewing out fumes. I was no doubt a forlorn-looking character, lost and confused in my new suit and tie in a city that at the time I found rather big and frightening. But then I got the faintest whiff of spring. I had no idea where it came from or how it reached my nostrils. And I can't even define to you now what the smell was. Just that it smelled of spring. A button went off somewhere deep inside and my perspective was altered. And that is what it is to hear a chaffinch sing in February.

Chaffinches can be seen all over the place. Originally they were a woodland bird but are now found in urban parks, suburban gardens, open farmland and the wildest of woolly places. It is thought there are six million nesting pairs in Britain. A good way to know if a chaffinch is in your vicinity, without seeing it or hearing its song, is the constant *weet, weet* noise it makes. In actual fact, the *weet, weet* is his alarm call, but old country folk took that as a sign it was about to rain, and many called the chaffinch 'the wet bird' as a result.

And if you are lucky enough to stumble across a chaffinch's nest, usually to be found in the fork of a branch in a tree or hedgerow, then even better. Of all the birds I think the chaffinch's nest is probably the one that is most perfectly formed (the song thrush comes a close second). It is a product of stunning workmanship, a Norman Foster design. The nest is completely circular – shaped like a neat cup – and made up of moss, grass, wool and feathers, decorated on the outside with lichen and held together by spiders' webs. If I didn't know a chaffinch had built it, then I would put its construction down to the fairies.

Another fine finch, and one that always conjures up warm June days lazing in the garden, is the greenfinch. It's a similar shape to the bullfinch and chaffinch but draped in a golden green smock with gold wingtips. When I first started looking out for them I often got confused because I was convinced I was seeing an entirely gold bird. But it's not gold; it's nature's most delicate shade of golden green.

The best time to listen out for a greenfinch is summertime in your garden. You will hear this persistent lazy *churrrrr* emanating from a distant tree, followed by the most magical string of light, clear notes. It really is a treat. It is impossible to be angry or stressed when you are bathing your ears in the greenfinch's sweet melody.

I've learned that by absorbing the sights and sounds of the wildlife around me, by watching nature – especially my window-bashing chaffinch – not to be ashamed of the anger inside. It is part of me. Innate. I no longer fight it or try to crush it. The trick, I've learned, is to come to terms with it: to redirect the emotion to more constructive channels. This is, of course, much easier said than done. And I still fail, regularly. I still curse inanimate objects, bellow at bad drivers, shout at my children, and undertake long, fruitless rows with my own reflection. But I've learned that there is simply no point in getting angry with my father – and his habits – because it is utterly pointless; I cannot change him. I'm at peace with that now. I focus on his good points, of which there are surprisingly many.

Years later he is back – via a few ups and downs – to his old self and we enjoy one another's company. He is once again that busily intent bullfinch. It took about twelve months before the first small shoots of his mental recovery began to emerge. The calls to his children diminished, as did his anger, and he regained the confidence to go horse racing on his own and to take up again slowly but surely his old hobbies and occupations. He even reached a point where he could conduct an entire conversation with a friend in the pub without mentioning Mum or his own misfortune.

And then, one bright, life-changing day, he found his Suzie. And all our lives changed again forever. Suzie is the second love of Dad's life, and the only other woman in the world with the patience to put up with him. We call her Saint Suzie, and, as Dad says to his children with boring regularity, 'You're very lucky I've got Suzie!' He's right, we know.

My children adore their grandfather (except when he is bellowing about lost dogs, upturned water troughs or open gates). They look forward to visiting his farm, which we do now most weeks, and, I hope, they are forming a special affinity with his sheep. Quink is now an old dog but remains devoted to my father. They are each other's living link to my mother.

My relationship with my father remains prone to volatility. Nobody is perfect. I'd be more concerned if we didn't have the occasional blow-up. Ask Mary. She'll definitely tell you I'm still learning.

'Just go for a walk and cool down,' she'll say when I lose control of my volcano.

'Good idea, my darling. I'll go and track down those bullfinches in the lane.'

My father has his sheep and dogs and I have the birds. The same but different.

Bullfinch

What it looks like: *A stocky finch with a 'bull head', black cap, stubby beak and grey back. The most distinctive aspect of the bullfinch, though, is its vivid blood-red breast. They also have a bright white rump, which is easy to spot as they fly off. Bullfinches go everywhere in pairs so if you see a male, you'll soon spot the female too – which is a duller, washed-out buffish-pink version of the more resplendent male.*

What it sounds like: *A gentle piping call that emanates from deep within hedgerows.*

Where to find it: *The bullfinch is a very secretive creature that likes to hide and nest in thick undergrowth, which is why you are far more likely to hear it than to see it. The best time to see bullfinches is in early spring, when they are feasting on the buds of trees and shrubs.*

What it eats: *It's very partial to the buds of trees and shrubs in early spring, which is why they were once known as the budding bird, and cursed by fruit farmers for damaging orchards. They also eat seeds and during nesting time catch insects to feed their young.*

Chance of seeing one: *This is a shy bird, so if you are walking in woodland near thick undergrowth in a rural area, you've only got about a 20% chance of seeing one. You are much more likely to hear one. Look out for them when the buds are out.*

6. Magpie

Seven for a secret that's never been told.
                    Old English nursery rhyme

Over the land freckled with snow half-thawed
The speculating rooks at their nests cawed
And saw from elm-tops, delicate as flower of grass,
What we below could not see, winter pass.
                    Edward Thomas

For two weeks I had watched with unbridled delight as two busy little mistle thrushes built their raggedy nest right in front of my sitting-room window. In the teeth of March gales, punishing rainstorms and primrose-crushing late frosts, these thrushes had never ceased. They had created their untidy nest out of anything to hand – bits of plastic, old carpet and even what looked like old unravelled cassette tape. The construction of this nest had been one of the great sources of solace for me during the 2020 coronavirus lockdown – a useful and optimistic role model – and had acted as a daily inspiration for my family and me. The mistle thrushes were singing through the storm and showing us all the meaning of perseverance in the face of adversity. I had shown my two boys the location of the nest, high up in the fork of a branch of a sycamore tree, and they had – to my great surprise and deep satisfaction – become fascinated by this miracle of nature, taking place just yards from their back door. We would watch them from afar, often with binoculars, so as not to give away the location of their nest.

And then the magpies came.

One late-April morning, after an unseasonably windy spell, I went out to check if my thrushes had made it through the night. By this stage the nest was complete and I was pretty sure Mrs Mistle Thrush was sitting on her eggs. I could usually tell quite easily whether all was well, because Mr Mistle Thrush would be singing loudly from the top of a nearby rowan tree from about 5 a.m. onwards. But this morning he was silent. It was all too quiet, in fact. I had a bad feeling.

And then I heard the harsh rattle of the magpie. I looked up and to my great dismay saw two of these black-and-white scoundrels diving into the sycamore tree by turns, as if laser-guided towards the vulnerable nest. The male mistle thrush fought them off with consummate bravery – belting out his furious war cry and chasing them off whenever they came close to his small family. But it was in vain. Magpies are tenacious and intelligent. A nest, once located, has little hope.

About twenty minutes later I inspected the scene of the crime. Beneath the sycamore lay large chunks of the nest – egg-stained – and above it, what was left sat forlornly askew in the tree like a ship run aground. I could feel the life had gone. The whole tree felt dead. And I didn't see the mistle thrushes again. But I now see and hear their persecutors every day. And I won't deny it takes some getting used to. They strut proprietorially around my garden like a couple of fascists at a rally. As I write these words I can hear that death rattle of the magpie outside my window. It pierces my skin, gets into my soul, and a small part of my inside tightens with despair for my thrushes. And I cannot help but draw a comparison in my mind to the cancer that had raided my family and took away my mother all those years ago, leaving our nest broken and askew. I constantly have to remind myself that nature is *not* sentimental, and neither is life. Where there is light, there is darkness. You cannot have one without the other.

A very wary, crafty bird . . . bold, impudent, thievish rascal. I might as well stop writing now. I'm not sure I can do better than the Rev. J. C. Atkinson, who in 1861 described the magpie in this way in his book *British Birds' Eggs and Nests* (price one shilling). I've also seen the magpie described in old texts as having *a natural propensity towards cunning and plunder . . . liable to be called to account for petty larceny at least, if not open robbery.* They didn't hold back, those Victorian bird enthusiasts. And after having witnessed the wholesale destruction of my beloved mistle thrush nest I cannot help but sympathize with the Rev. J. C. Atkinson and his nineteenth-century peers.

Its name, 'magpie', literally means chattering (*mag*) black-and-white bird (*pie*), and it is true to say that it has never been a hugely popular creature through the ages for entirely rational as well as irrational reasons. Rational because, like their cousins the jackdaws, crows and jays, magpies steal the eggs and newly born young from the nests of other songbirds (the mistle-thrush-murdering beasts!) and irrational because humans have always attached to them no end of baseless superstition and fearful folklore.

A single magpie has always been seen as a bad omen: *one for sorrow*, as the nursery rhyme tells us. One of the most enduring memories I have of my childhood is my mother imploring us to chant 'Morning, Mr Magpie, how's your wife?' whenever we saw one loitering nearby to ward off the evil spirits (while Dad would tip his greasy old hat). If we didn't, heaven knows what would become of us. Though on the upside, as the rhyme also taught us, it was two magpies for joy, three for a girl, four for a boy, five for silver, six for gold and seven for a secret that's never been told. So, all in all, and as long as you tipped your hat or asked after the health of his wife, seeing a few magpies on the verge as we speeded off to school was an altogether positive experience.

But I've still always associated magpies with bad luck. And I won't deny that seeing one outside my bedroom window around the time of Mum's cancer diagnosis set off all sorts of irrational thought processes. Reason deserted me, because, after all, in Scotland a magpie seen near a window is a sign of impending death. (Not that I was in Scotland; I was in Wiltshire, but Mum was half Scottish, so it counts.) In Wales it is terribly bad luck to travel if you see a magpie. If you're a Devon fisherman, you won't catch any fish that day if you see one in the morning, and if you are unfortunate enough to see three magpies as a resident of Northampton, then your house will burn down.

'All this endless superstition must be utterly exhausting for you,' Mary often says, exasperated by my constant doffing and singing and walking backwards in circles. (It's not just magpies that set me off.) And she's right. If brought to its natural conclusion, it's probably safer never to leave the house, just on the off-chance you see a magpie and someone dies, you have a car crash on your way to your fishing boat, but then you wouldn't have caught any fish anyway, and, besides that, your house will probably have burned down in the interim. Though you can always follow the example of the people of old Somerset, who wore onions round their necks as a sensible precaution to ward off the evil of any potential magpie encounter. Sound thinking to my superstitious mind.

It's not just in Britain that these irrational superstitions hold sway. All over the world people fear and respect magpies in equal measure. In many other cultures and times they were revered. The Romans were big fans and saw the magpie as a creature of great intelligence and reason. While in Native American folklore magpies were regarded as a sacred messenger, and to wear a magpie's feather was a sign of fearlessness.

It never ceases to amaze me how deeply engrained the birds are into the human psyche, hardwired into our culture, and what a powerful force they have been in the world's shared history. While I regard magpies as my enemy, both in terms of my irrational superstition and entirely rational fear for the future of my garden's mistle thrush population, none of this is particularly fair on the magpie. After all, a magpie doesn't know it's a magpie. It just does its magpie thing.

In truth, and with my rational hat on, the magpie is a beautiful creature with its elegant long tail and tidy white-and-black uniform. They are charismatic and tenacious birds with a fine intellect. In medieval times people often kept them as pets, before the priests explained to them that magpies had a touch of the Devil's blood and could never be trusted – those Jesus-betraying fiends. It is said that the magpie, as well as the wren, betrayed Jesus to the Romans in the Garden of Gethsemane. I'm not sure what it was he did to upset all these birds.

And while I will publicly condemn the magpies around me to friends and family alike, it is impossible to hate a magpie. They are nature's necessary adversaries. And in life we all need a necessary adversary or two. It's good for the soul: black and white, darkness and light.

The magpie is part of what is known as the corvid family of birds. These include ravens, rooks, crows, jackdaws and jays. Like the magpies, these birds have always been associated with death and sorrow. No horror movie or tragic funereal scene is complete without the obligatory solitary crow cawing eerily in a nearby tree or rooks circling an ancient beech – wings outspread – towering above a fog-filled graveyard. A picture of doom. I get a chill just thinking about it. But, in actual fact, corvids are cheerful and gregarious animals that exude life not death. Hearing the sound of a clattering of jackdaws, chack-chack-chackling away on a cool autumnal evening always gives me pause for a happy reflection or two. Jackdaws are so black they're almost purple and have an iridescent silvery shimmer to their colouring that is beautiful to behold. They're a smaller, more mischievous, chuckling, chatty version of a crow, and very easy to spot because they hang out in large mobs in town and country alike, and nest in nooks and crannies all over people's houses – and sometimes, much to my personal annoyance, in my chimney. If Danny DeVito were a bird, he'd be a jackdaw.

And if Danny DeVito had an extrovert cousin who liked to dress in garish clothes, then that would be the jay. The jay is another characterful member of the corvid clan. Though instead of the usual sombre black ensemble, the jay sports a pink suit, bright white shirt with dazzling electric-blue wingtips and a snappy black moustache under the beak. You'll certainly hear a jay before you see one (its Latin name is *Garrulus glandarius*). If you are walking through woodland, your ears will be assaulted by a shattering shriek and, just as you remark to your walking companion, 'What the hell was that bloody awful noise?', you'll see this pinky-blue-white blur fly past with a kind of lazy undulating lollop. Yet despite its shouty call and garish costume, the jay is, in fact, really rather a shy creature. It lives in woodland and rarely ventures out.

The most attractive quality of jays, though, is that they

are nature's forester. And while the jay is a keen and dedicated nest raider, and so will never be a close friend of mine, it is also responsible for planting out more woodland than the Forestry Commission; they are the original environmentalists. Jays collect acorns all summer long and bury them in different places all around their territories to come back to when winter comes. They can carry up to nine acorns at a time, and some say they can bury up to 5,000 in a single season. I'm not sure what is more astonishing: the fact that jays have time to bury that many acorns or the remarkable feat of memory magic it must take to remember the precise location of each one. I can't even remember where I left my car keys this morning, let alone find an acorn I buried two months ago in a wood a mile away. Although they clearly don't find all the acorns they bury or we wouldn't have any oak trees.

Of all the corvids I think it is the rooks that intrigue and beguile me the most. As a child growing up, we had two huge rookeries just up the hill from us. Part of the soundtrack of my childhood is the cheerful racket of those rooks, although they're not popular with everyone. If you're fond of the sound of silence, then moving next door to a rookery is not for you. And I have sympathy with the sentiment that it is much easier to enjoy the pleasure of watching a rookery outside someone else's bedroom window, rather than your own.

Rooks are extraordinary creatures, with their deep black sheen and grey-white beaks. They live in loud, clattery communes, the aforementioned rookeries, and can be found all over Europe in towns and cities, but most especially on farms and farmland. The best way to tell them apart from their close cousin the crow is that rooks will always travel about in big flocks, whereas crows tend to fly solo. Or, as the old sages used to say, a crow in a crowd is a rook and a rook on its own is a crow.

Rooks will find a tall beech tree, or ash or oak or sycamore – something nice and thick – and take complete possession of it, building their ungainly, ramshackle nests right at the crown. Some of these rookeries can dominate five or six trees and contain 200 nests or more – like having a football crowd nesting in your garden.

The more senior you are in the rook hierarchy, the higher up the tree you can build your nest, so they say. They are monogamous, and the young always return to the nest where they were born. In fact, rooks have highly complex social rules. Naturalists have been studying them for centuries and still admit to having little understanding of their rook-based codes of conduct, though I shouldn't think they're too far off our own society's rules and prejudices.

Rooks even have courts of justice. Stories abound in history of sightings of large numbers of rooks gathering in circles around a single rook as if in judgement upon it. People claim that if convicted of a crime against rook society, the guilty parties will be pecked to death. Not having seen one of these courts of justice in action I cannot speak to the veracity of the tale, but what is certain is that the collective noun for rooks is a 'parliament', and this derives from these supposed courts of rook justice.

But it is the magpie, and its nest, rather than any other of the corvid clan, that draws me inexorably towards meditations on Mum and the strange event that took place on her birthday one month after she died. Magpies' nests are known for their inaccessibility; they are large domed affairs perched high up on their own in trees. These nests are formed of vast quantities of sharp thorny sticks and when finished resemble a large ball of spines with only a small hole left undefended, from where the magpie can watch warily for signs of unwanted intruders. And it was on this day – what would have been her birthday – that I saw a parallel between those unreachable magpie nests and my mother's own approach to life. I saw for the very first time the pain-filled darkness behind all Mum's light.

We had gathered as a family to mark the event with a dinner at our childhood home. Mum's younger brother, Andrew, and his wife, Sue, had joined us that night. We'd eaten dinner and settled ourselves in the sitting room in front of a crackling fire. My father, draped in his usual chair, my sister kneeling somewhere near his feet, Mary and I perched on a nearby sofa, and Uncle Andrew propped up by the fireplace, next to my brother, drink in hand.

We were all tired, a bit drunk, and the talk was of happy memories of Mum and how much we missed her. You could smell the woodsmoke and sweet sorrow in the air. And then somebody, I think my brother, mentioned *the box*.

The box was a small nondescript wooden chest – the sort of thing you might keep jewellery in – that had sat on the windowsill above Mum's dressing table all my life. It had a (not so) secret key that was hidden behind a sliding panel on the front. There was only one rule with the box when we were children and that was, do not open the box. You could pick it up, carry it to and fro, shake it and even take the key out of its secret hiding place – all of which my siblings and I did on numerous occasions – *but we never opened the box*. Ever. Even though, it later transpired, each of us had thought that one or the other of us had indeed opened the box but kept it a secret. All those secrets!

That night, though, we opened the box. At first we were all so fearful of what might befall us, what we might find, that nobody wanted to open it. We sat round it like a parliament of hesitant rooks. I think there was a lingering fear that Mum might walk through the door unexpectedly and furiously snatch it out of our hands. It was Uncle Andrew who finally broke the stalemate. Since none of Mum's children had the courage to open it, he took it off us and did the job himself. He'd never before heard of this box, but I sensed that he knew very well what was inside it, as did Dad, who was inspecting the scene from the safety of his comfortable armchair, muttering in vain, 'Mum wouldn't have wanted you to open the box!'

We disagreed. And so the box was opened. And all Mum's secrets fell out on to the carpet: unwanted truths scattered in a shower of yellowing letters and postcards.

We all knew that the box had something to do with Mum's elder brother, Michael, whom she had adored, but who had died in his early twenties before any of us were born. Mum never spoke of Michael except to say that he had died in some kind of gas-related accident. A story I had had absolutely no reason to doubt. But the box told us another story.

Among cheerful letters and postcards to Mum from Michael, loving keepsakes, were other letters: letters of condolence written to Mum after his death. Letters that spoke of a 'disgraceful verdict' of 'disbelief' and of 'shock'. Letters that contained phrases like 'it could not have been suicide'.

The unvarnished truth, it transpired that night, the night of Mum's birthday, was that after Michael had died the coroner had recorded an open verdict. *He had refused to rule out suicide.* Three years after Michael's death their father, aged just fifty-seven, had collapsed and died of an aneurism – or maybe it was a broken heart? Because what we also learned that night was that a letter from Michael had been found in his wallet, which he had evidently carried with him ever since his son had died.

'What did the letter say, Uncle Andrew?' we asked in horror-struck unison.

'It ran to several pages and said how unhappy he was,' he said.

'Your mother and I burned it.'

I'm not sure what is worse, knowing for sure that a brother you adored took his own life or not knowing for sure whether a brother you adored took his own life, having that stain hang over your memory of him like a heavy coal-black and freezing February rain cloud. And so Mum had buried it in her wooden box. She locked it up deep inside and never spoke of it again. Yet this small wooden chest full of so much pain had sat in plain view on her dressing table for the rest of her life. A daily reminder of her loss. To this day I cannot fathom it.

Our mother's childhood ended the day her brother died. She was nineteen. On the very rare occasions Michael was resurrected in conversation she shut it down with an iron full stop and moved hurriedly on. She toed the line; he died in a gas leak and that was the end of the story. My late uncle and grandfather were ghosts in my childhood: shadowy, faceless figures that were part of somebody else's history and not my own (something I am working very hard to put right). I used to scavenge crow-like for scraps of their history: a brave war story here – my grandfather had been in the Secret Intelligence Service in the Second World War (more secrets) – or the odd racing driver anecdote there (Michael used to race cars apparently), but I learned very little. I never got inside that magpie's nest. I didn't find out my grandfather's name or see a picture of him until I was in my teens. It was the same with Michael.

And that is where Mum wanted them, buried so deep they would never emerge again. She took the same attitude about all her childhood; she rarely, if ever, spoke of it. The pain of its sudden destruction must have been too much for her to bear. It was like her life had started in 1968, the year she married my father – one week after the death of her own father.

Not long before the end, I was on a night shift with Mum. I've no idea what time of the night or morning it was, but I'd been reading out P. G. Wodehouse novels to her – a favourite childhood author of Mum's, and of mine, too. In fact, it was Mum who had introduced Plum into my life. And what a lifesaver he's been over the years. I rarely travel anywhere without an old battered copy of P. G. W. by my side. He is an oasis of good humour and reassurance no matter how far from home I am, how sterile the hotel or how bleak my circumstances. He's the literary version of my birds. Anyhow, Plum was a real connection between Mum and me, and, I'll be honest, one of the few genuine interests we shared.

While Mum dished out love and human kindness like free ice cream on a hot day, and though – or perhaps because – she put everyone else's needs and dreams above her own, she was hard to reach. One never really had the feeling you knew quite what was going on beneath those shining green eyes and the warm smiles. Her love was real and tangible, that much we could be certain of, but did we ever know the real Mum? She built a magpie's nest round her heart and kept her secrets deep inside, behind a thick ball of metaphorical spines. Any attempt by me, or any of her children, to unravel this nest was met with fierce resistance, and she kept those secrets from her children till her dying day.

As a result, my relationship with Mum was complex and hard to unravel. To this day I feel I've never been able to get to the essence of Mum. The recipe remains a mystery. So that is why our shared love of P. G. Wodehouse was hugely important to me; it was ours and ours alone. My father ('I've never read a book!') and siblings are not big readers, but I've always liked nothing better than to bury myself in a book. I am an escape artist from life, and I'm not ashamed to admit it.

I wasn't entirely sure if she was registering or not, as I recited tales of Wooster and Jeeves, Blandings Castle and Mr Mulliner, but I believed on some level that P. G. W. was filtering through. At one point she woke up, and was quite coherent, too. I was overwhelmed that I could finally conduct a half-decent conversation with her, after days when she had been too ill really to talk. I held her hand, and foolishly, against my hardest endeavours, I started to cry. It just sneaked up on me without warning. When I should have been administering hope, I wept like a child. I could see how much it upset her to see me sobbing away, but I just couldn't help it. My overwhelming feeling was that if she saw me crying she might think that I thought she was going to die. She looked so worried. She couldn't bear to see me like that and wanted to console me. 'Please don't cry, Charlie,' she said over and again. And those are some of the last words she ever said to me.

The revelations about Mum's hidden past hit me like an emotional cement mixer in the ribs. It hit all her children in this way. I truly had had no idea – not even the smallest suspicion – that anything so truly awful had happened in her life. Uncle Andrew talked movingly of their childhood, of Mum and their late father and brother. In some ways it was like he was talking about someone else, an unrelated family, and not our mother or my relations. It was like someone, at some point, had disconnected me from my history and I had finally been plugged in again.

So that night I started to fill out those ghosts of my childhood with personality and stories, to get a sense of who these once faceless people, my direct relations, were. And I continue to this day to build them up in my mind. So I might get to know and to love them like my mother must have done.

In one of the very few conversations I had with Mum about Michael, in my early twenties, I remember her so clearly saying that not a day had passed since he had died when she had not thought of him. It was Michael who had passed on his love of P. G. Wodehouse to Mum, and she had passed that on to me. I feel this bond very strongly, even though I never met him. It is, after all, all I have. Recently Uncle Andrew produced hundreds of pictures of their childhood with Michael, pictures I had never before seen. Image after image of this happy, smiling boy growing up into a happy, smiling young man. I could see in him strong flashes of my brother, my nephew and my own son, Arthur, and I wished so powerfully that I could have had the chance to meet him.

Like an angry rook, I sat in judgement upon Mum for quite some time after the revelations. Why had she never opened up? I understood the desire to bury the pain, of course, but did this have to mean so completely severing my siblings and me from our roots? I still find it hard to get my head around. I make sure that my sons, who never met Mum, are surrounded by pictures of her and are regularly regaled with stories and memories. I want them to feel on some level that they *knew* her. My eldest son, Arthur, for example, knows for an absolute fact that he has his grandmother's smile; I tell him every day. Why could I not learn about my grandfather or uncle in the same way? But I suppose I live in a much altered, more open age.

In Mum's upbringing you never complained and you never explained. Grief and loss were your own business and it was rude to burden others. Instead you built a thick nest of thorns round your heart and carried on. Mum was forever quoting Winston Churchill's maxims 'Keep buggering on' and 'If you're going through hell, keep going'. Now I know why. She cancelled the pain forcibly and devoted her life to making damn sure her children, and all those around her she loved, lived in an atmosphere of happiness, security and comfort. And while Mum never spoke of her past, we all knew one inalienable fact as children, that whatever happened *we were not allowed to die*. Mum might have kept her secrets inside her magpie's nest of a heart but it was camouflaged by love.

This morning, as I write these words, eight years after that difficult and confusing night, the robins sing and the rooks gently caw. The jackdaws chack and the magpies rattle. There is a mist that feels like it is rising from beneath the ground. It breaks apart the weak sunbeams and gives my garden an ethereal air. The field maple in the paddock beyond my house spreads itself in the morning light, shaking off the droplets of mist, and stretches out as though waking from a long sleep. Young cows graze contentedly beneath it. And somewhere, close by, someone has been kind enough to light a bonfire. Its rich smoky aroma completes my autumnal fantasy.

This is a single minute of a single morning, in a single autumn in a single year. And there is nothing else in the world I want more and nowhere I would rather be, right now, than amid the sights, sounds and smells of this season and this morning. And while I accept that very many of my autumn mornings will be drizzle-soaked, cold and grey, with not even a friendly bonfire's aroma to massage the senses, the mist-soaked and sunbeam-filled days are never far away. They will return.

When I close my eyes, I can see a similar day of my childhood. The beginning of a new term, the school run, with Mum in her sky-blue Ford Escort (called Oscar) racing far too fast down the lanes of my childhood. A pair of magpies is startled (no doubt by Mum's attempt to mimic Niki Lauder) and she cries out cheerfully, 'Morning, Mr Magpie, how's you wife? . . . Two for joy!'

Sorrow and joy. Darkness and light. I can't have one without the other. They are necessary adversaries.

Magpie

What it looks like: *A medium-sized member of the crow family of birds, a corvid, with a black back and bright white belly. The magpie's long dangling tail makes it very easy to spot in flight, and its wingtips have an iridescent blue-green sheen up close.*

What it sounds like: *A harsh, quite loud, chattering and rattling call.*

Where to find it: *The magpie can be found in abundance in town and country alike – in any park or garden, although there are fewer magpies in upland areas. It likes to build its thorny nest at the crowns of dense trees and shrubs, and will often be seen loitering at the side of busy roads.*

What it eats: *Magpies are scavengers. They eat anything from dead mammals, especially roadkill, to earthworms, fruit, insects and discarded food. They are most notorious, though, for raiding the nests of songbirds in the springtime, eating the eggs and sometimes young chicks too.*

Chance of seeing one: *You have a 95% chance of seeing this bird if you go on a walk in any town or city. They are remarkably adaptable birds thanks to their keen intellect and varied diet. But where you have lots of magpies, you will have fewer songbirds.*

7. House Sparrow

Only connect! That was the whole of her sermon. Only connect the prose and the passion, and both will be exalted, and human love will be seen at its highest. Live in fragments no longer.

E. M. Forster, *Howards End*

A sparrow's nest can be discovered without search but cannot be reached generally without trouble.

Edward Grey, *The Charm of Birds*

It is the old books about birds that I love the most. When our ancestors wrote about wildlife they wrote as though the bird in question were an old friend or age-old foe, and I get a very tangible sense of the familiar. After reading a description of a bird in one of the older dog-eared books I have hanging about the house, I really feel I've got to know it. Not just a basic knowledge of what the bird in question looks and sounds like – or where it lives – but its personality, too. It feels like part of the family. It feels alive when I read these ancient texts.

That is not to denigrate the modern writers of guide-books; it's just that their descriptions feel a little less familiar, a little more worthy and a little too scientific. They write about birds as something to look at, to study and make a note of, rather than as of an innate part of our daily life, of the human story; there doesn't seem to be any real affection. Take the humble house sparrow as an example. Here's an entry about the sparrow from C. A. Johns's book, *Birds' Nests*, published in 1854, and it makes me want to go outside, greet every sparrow in the neighbourhood and invite them in for a drink:

> This familiar, nay saucy little bird, would seem to have lived so long in the society of man, as to have learned to imitate him in the variety of his dwellings. No place comes amiss to it. Does a chimney smoke when we light our first autumnal fire? – Ten to one that a pair of these birds has not stuffed it up with a bundle of hay and feathers. Is a leaden spout found to be choked? – A house sparrow, in all probability is the culprit. Does a wooden shoot unexpectedly overflow during a thunderstorm? – No one is to blame but *those troublesome sparrows*.

The book goes on this vein for quite some time, including a heated passage about sparrows' habits of repossessing house martins' nests, labelling them *as dishonest as they are idle*, before finishing up: *But though the sparrow is impudent to men and overbearing to other birds, its character is redeemed by its extreme attachment to its young*. If that doesn't make you want to rush outside and meet an impudent sparrow, I don't know what will.

Now compare that elegant prose to my *Collins Complete Guide to British Birds'* entry for the sparrow, published in 2004: *A familiar species, in part because of its affinity for human habitation. The House Sparrow frequently dust-bathes and small groups are often encountered sitting on roofs, uttering familiar sparrow chirps . . . the sexes are dissimilar.*

The sexes are dissimilar. It almost feels like someone's written that deliberately to put people off. It's all cold hard fact and no heart.

In the year after Mum's death I knitted the birds into the fabric of my daily life. I began to treat them not as something to look out for and identify near 'human habitation' but as a kind of extended family. And it was the common house sparrows outside my door – so often ignored and disregarded – that I knitted myself tightest to for a time. Not just because they were the most abundant and relentlessly cheerful of all the birds in my garden, but also because they so closely resembled my own extended family: a noisy gregarious bunch of individuals, often squabbling (one collective noun for sparrows is a 'quarrel') but always around for each other when it mattered (another collective noun for sparrows is a 'ubiquity'). I had always taken for granted that whenever a common enemy reared its head – the sparrowhawk of fate – my family would flock together for protection. But Mum was gone now. The sparrowhawk had done its worst, and life needed to go on.

After all the togetherness during Mum's illness, and in the immediate aftermath of her death, we separated: Dad retreated to the farm to lick his wounds, Katie back to her family in Edinburgh, Richard to his family in Hampshire. And me back to Mary and my job in London. I attempted to carry on with life exactly as I had left off seven months before. I mean, what other choice did

I have? They always say routine is good for you after a tragic event. And it is, up to a point. What I hadn't bargained for, though – as I fluttered off cautiously into real life again – was quite how much I had been affected on a deeper level. There were consequences. And for years I did not link these consequences on any level to the death of my mother. I just saw them as personal weaknesses that were a source of great shame to me, that needed to be hidden from view, and that I could fix through my own strength of will.

The most obvious of these consequences was a heightened sense of anxiety. Well, I say 'heightened' when what I actually mean to say is 'almost crippling'. I had always been a worrying type – at times rational and at other times totally irrational – and so this escalation of anxiety at first went unnoticed. I didn't even realize it was anxiety. Anxiety to me had always been something you suffered before a football match or tricky exam – or waiting for a job interview. A big blinking neon sign – *ice on the road ahead* – not something that sneaks up on you from behind and when you least expect it knocks out your ankles with a hockey stick. One day this heightened sense of anxiety did just that; it announced itself quite spectacularly out of what seemed to be a clear blue sky. But looking back, I realize now that I was an accident waiting to happen.

For quite some time I had not really been enjoying my job. What I had in the past found creatively rewarding, I now found tediously formulaic. What I had in the past found exciting and life-affirming, I now found deeply stressful. People I had in the past found engaging and fun to be around, I now found to be a source of irritation. I didn't tell anyone, of course. I kept up my facade (or at least I think I did) of the cheery soul in the office, a decent bloke to share a bit of gossip with or have a go at management with – grumpy at times – but ever-ready for an affable lunch at El Vinos or a few drinks in Ye Olde Cheshire Cheese.

'I'm amazed,' one extremely friendly US colleague said to me one day. 'When my mom died I needed three months out of the office to recover, and you're back in a week.'

'Oh, well, you know,' I said out loud. And looked at the floor and felt embarrassed. Inside I thought, *Blimey, are you allowed to do that? Just take three months off?*

I was with my father one weekend (most weekends were spent making sure Dad had enough bananas and gin to keep him going, and the odd cottage pie). We'd gone together to a local race meeting and met up with a few good friends of his. He and Mum had been keen followers of the horses and were forever going racing together. As we stood in the March drizzle, discussing form and 'the going' and various other matters of the Turf, one of Dad's oldest friends pulled me aside.

'I realize how difficult your father is being these days. It must be very hard for you and Richard and Katie. How are *you* getting along with it all?'

It seemed a pretty innocuous question but it knocked me sideways. The stock question I had become used to being asked by the vast majority of people in the weeks and months after Mum's death was: 'How is your dad?'

For the very first time here was a man – and one of the last people I had expected to say it, too – who saw it as a tragedy for me and my siblings rather than just for Dad.

It was extremely disconcerting. And I didn't know what to do with myself. If I had been a child, it would have been one of those moments after a brutal day at school, when quite unexpectedly someone puts an arm round you and asks if you are OK, and you just burst into tears on the spot. I was completely disarmed. I didn't cry then, of course, as this old-school friend of Dad's – reared on cold showers, pipe smoke and whisky – comforted me. All I could think to do was to shuffle a bit on both feet, and say, 'Thanks, Fred, doing well. And thank you so much for asking.' And then I changed the subject.

I was back at work the next day. It was 10.30 a.m. and I was in our morning meeting. I was listening intently to a colleague talking earnestly about upcoming special reports and projected ad spend, or something else equally mundane and uneventful, and waiting for my turn to speak. All of a sudden I was gripped by an overwhelming desire to escape. Quite where this feeling came from I have no idea, but I needed to escape. Right now. I wanted to be anywhere but this room. But what would people think of me if I just ran out of the room? That was unthinkable. I was trapped. Trapped. And, Christ, I needed the loo. But I couldn't escape. Oh God, Oh, God. Oh. Dear. God. I couldn't breathe. The room shrank. The walls and the people came in at me, and it got very, very hot. My cheeks felt like they were on fire. The windows were all sealed shut and I was as far as it was possible to be from the door. Wedged between two people at the back of the room with my back to the window. The door was firmly, horrifically closed. I needed air so desperately. My head was at the centre of an explosion and my world was collapsing in on me. And yet it was just another drab Monday morning at the office. My life was imploding – I.M.P.L.O.D.I.N.G. – in an obscure little meeting room on the fifth floor of an office block near Ludgate Circus, with a cup of cold coffee in my hand and a small Danish pastry on a plate in front of me.

It made no sense whatsoever. And the reason it made no sense at all was because in the past I had always prided myself on my ability to speak publicly. I was confident in front of friends and strangers alike. I would even go so far as to say I was a little too pleased with the sound of my own voice. Always the first to volunteer to give a presentation, interview a bigwig or talk on the radio about the impact of inflation. I was regarded as a good and entertaining speaker in the public sphere. And I enjoyed it too.

'Charlie? What do you think? Charlie? Are you OK?'

'Erm, what? Yes, yes, thank you. All good!'

What to do, what to bloody do? *Oh, Christ. Can't let them think I'm about to soil my trousers. Oh dear Lord, what to do? I can't escape. I can't escape because running out of the room would be just as humiliating as having an accident. Oh God, Oh God, Oh God. Drink. I'll drink lots of water.* And so I started to gulp water, but then thought, *No! This will make me wet myself.* And I couldn't move either. I felt that if I moved at all, even slightly, I would have some dreadful accident.

In my fevered state I resolved to talk and to talk a lot. Very fast. It seemed the only way out.

Talking seemed to help. As my words sprayed out like machine-gun bullets, my mind seemed to ease a bit. Heaven knows what I talked about. I had no recollection even five minutes later. But I got through the ordeal. Just about. And I don't think anyone really noticed. Maybe they thought I was just a bit hungover.

At the end of the meeting I rushed to the bathroom (it turned out I didn't need the loo after all) and stuck my head under the tap. Still shaking, I took the lift down to the ground floor and crashed out into the street. I took huge gulps of diesel-tinged air, followed by lungfuls of Silk Cut, and thought, *What the FLYING FUCK was all that about?*

I know now, of course, that what I'd just been through was a grade-one, bona fide, all-expenses-paid panic attack. But at the time I was nonplussed. I walked up to a little hidden square I knew tucked away behind Ludgate Circus – not pretty, more an open-air concrete bunker, and usually inhabited by drunks after work spilling out from the soulless chain pub in the corner. But I wanted to go there because you sometimes got a glimpse of a peregrine falcon flying about. A pair nested somewhere nearby. And one spring day, a year or so before, I'd even seen two grey wagtails mobbing a falcon. That's like two biplanes attacking a Tornado jet. Yet these spirited little birds had remarkably succeeded in seeing him off. I was ever hopeful of catching a glimpse of the peregrine, not least because seeing one of these creatures in action is one of the greatest free shows open to Londoners, if only they would look up occasionally. While the grey wagtails had succeeded in seeing off the peregrine that day, they were very lucky that they hadn't been taken by surprise. When this exquisite and well-proportioned medium-sized bird of prey, with its slate-grey back and white breast, sees a tasty-looking bird from above, it will lock its wings and hurtle through the atmosphere at 200-plus miles per hour towards its helpless victim. And the allusion I made to the Tornado really isn't far off; jet designers used the shape of the peregrine's nostrils – which protect the bird's lungs at high speeds – as a means to increase the safety and efficiency of their engines.

The peregrine is also one of London's great bird success stories. Having been largely wiped out by the Ministry of Defence by the end of the Second World War, due to their predilection for eating carrier pigeons, the population of breeding peregrines in the City of London is now thought to stand at thirty pairs. They like nesting on the ledges of the tall buildings and eating London's boundless supply of feral pigeons.

But I didn't see a peregrine that day. I just sat on a concrete bench and tried to work out what had just happened. I was mystified and frightened. What if it happened again? And that thought – *What if it happens again?* – became my downfall. From that day forward every time I entered a meeting room I thought *What if it happens again?* And then it did. Soon it wasn't just meetings; it was in all public places where a door slammed shut behind me. Barbers, cafes, shops. Church services for some reason became a particular problem, which wasn't ideal since so many of my friends were getting married at that time in my life. I spent a lot of time hanging out in graveyards, smoking.

I was very bad in the mornings. And so, quite systematically, I sliced out any meeting that might take place in the morning or just after lunch. I ended up with two hours a day, between 3 and 5 p.m., where I felt confident enough to get through a meeting without one of these attacks. In the end I stopped attending meetings altogether; I became an expert in making excuses or dialling in to meetings. Although strangely I would also have these attacks while on the phone, sitting at the safety of my desk. It was utterly and completely debilitating. I would not go anywhere unless I knew I was within a few easy feet of a lavatory. It made car journeys hellish.

It didn't just manifest itself in the form of individual moments of terror. There was a constant voice in the back of my head, a demon radio station whining on and on, telling me all the things that could go wrong in my life, from the mundane to the ridiculous. From not saving enough for my pension and dying in penury to horrific car accidents and a lingering death. My imagination was in overdrive and my brain became fearfully adept at making up extraordinary catastrophes that would befall me and the ones I loved. It never ceased, this whining little voice. As Mum would have said, had she been alive: 'It's enough to give your arsehole earache, Charlie!'

Some days I felt as if I wouldn't be able to get through the next few hours, let alone the next day. I could be sitting at my brother's house, around us a happy hub-bub: children's voices, laughter and the vibrant clatter of daily life. And there I would be, smiling away and chatting to Richard or his wife, Bebe, but quivering with anxiety inside, suffering from the onset of an unnamed dread. What if it all goes wrong? What if? What if? What if? These thoughts racing around my brain over and over. Funnily enough I am at my most imaginative and creative when thinking up new and tragic ways that my life could be changed forever and for the worse.

I saw it as a perpetual struggle against happiness. As soon as I scented even the most fleeting glimpse of happiness bubbling up inside me, I fought against it. I believed that if I accepted happiness, if I welcomed it to my breast, then fate would immediately deal me a smashing blow and snatch it away again. And yet if you were to know me at this time, you would have had no idea all this was happening. I wore the cheerful mask and went about my business.

One day I read an article in the Sunday papers about a technique called cognitive behavioural therapy. It sounded like something you might send a toddler to who was biting people in nursery, but the more I read, the more I liked the idea of it. The premise of CBT is that it deals with problems in the here and now, rather than nosily delving into the past to find solutions. I really didn't want to be told by some smug therapist – who didn't know me from Adam – that it was all my parents' fault or something to do with English boarding-school life in the mid-1980s.

I had never even considered a therapist before, and I had a feeling of profound shame that I felt I needed one now. It was a sign of weakness. And I was deeply furtive about the whole process. I didn't tell a soul, not even Mary. In fact, I'd have been less embarrassed about being caught going into a seedy Soho sex shop. But I screwed my courage to the sticking place and ploughed on. Something had to be done. I simply could not go on like this. It was not so much that I could no longer do my job, but my entire world was shrinking. Pretty soon I'd be unable to leave the house.

Eventually I located a CBT specialist near Victoria station in London, not too far from my office. He was a terribly nice chap. 'American, you know? But really *very* nice,' as my granny would have said. And that is how I treated him. Over the course of four one-hour sessions I was excessively polite while I guarded my secrets with an iron will. I gave nothing away. Not even the surging panic I was desperately trying to control, right there and then, as I sat not two feet from him in his tiny little breathless studio. I just nattered on like I was talking to a friend of a friend at a drinks party in 1958. And all for £80 an hour.

I felt like there was no cure at all for someone like me. I had battled so hard through those sessions not to give anything away about myself. And yet this anxiety was real, and it was crippling me.

My one lifeline was Mary. Her support at this time was quiet, unflinching and patient – with a heavy emphasis on patient. My exhaustion manifested itself in two ways: physical exhaustion and mental exhaustion. I could handle each of these taken individually, just about, but when they coincided all hell would break loose. And Mary was on the frontline.

In normal times Mary and I complement each other quite well. Her busy-as-a-bee nature and practical 'senior prefect' approach to forward-planning acts is an excellent foil to my general procrastination and let's-cross-that-bridge-when-we-come-to-it, path-of-least-resistance attitude to life. But yin doesn't always get on with yang. At times yin gets fed up with yang and vice versa. It's the busyness aspect of Mary's life that really sets me off, especially when mental and physical exhaustion combine, as they did then. (And still do at times, with two boys in the house under five years old.)

Mary is the busiest person I know. In fact, she is the busiest person anyone I know, knows. From the moment she wakes in the morning to the moment she runs out of batteries at night, head face down in the pillow with her hands still clasped to the iPhone, Mary is busy. Work, children, family, extended family, friends, house alterations – her own problems, her family's problems, friends' problems, friends of friends' problems. Fuelled by an overwhelming desire to help people, Mary likes to be involved in the endlessly fascinating business of life at every level. Sometimes I feel exhausted just looking at her.

I'm not exactly a later riser but by the time I crawl into the kitchen at about 7.15 a.m., one eyelid still glued down, slippers on backwards, gasping and grasping for tea, Mary has unpacked the dishwasher, cleaned the kitchen, sent three work emails, redesigned the sitting room (twice), created two cardboard ninjas for the boys and produced a spreadsheet that neatly summarizes the savings we need to make over five years if we want to send them to private school.

In comparison my single achievement that morning will have been to make it from the bedroom to the kitchen without stubbing my toe. My brain will just be easing into first gear, while Mary's will already be in fifth. And there will be questions: about the children, the house, next year's holidays, work, wallpaper design and weekend plans for the Saturday after next. The kinds of questions I really am not ready to answer until I've drunk two cups of tea – and one cup of strong coffee – stared blankly into space for fifteen minutes, and then eaten a banana.

And while one of the biggest reasons I love Mary is the fact that she challenges my strong opinions and is unafraid to deliver news about my behaviour that I don't particularly want to hear – and which no one else would tell me – at times on these mornings the wires in my head get crossed and I will verbally lash out at her. I'll become utterly convinced of some (completely im- agined) slight on my character, or simply get cross that she doesn't understand – or seem to care at all! – about the fact that I cannot find the key to the shed. Not that I need to go to the shed, but I might do later. 'Why can you never put keys back where they live?! Now I can't lock it, and we'll get robbed.'

'Don't shout at me.'

'I'm not shouting at you!'

'You *are* shouting at me. Stop it. You're turning into your father.'

I'M NOT BLOODY TURNING INTO MY FATHER!

'OK, so I admit I just shouted then, but I wasn't shouting before. But the key –'

Mary will look at me, baffled, and say, 'The problem is, Charlie, you're just arguing with yourself. You've created this row out of thin air – and for no reason.'

Infuriatingly I always know her to be right. And later I'll realize that I left the key to the shed in the door of the shed when I popped in there last night. But rather than dent my pride, in the weeks and months after Mum died especially, I storm off to work under a thundery black cloud and make a point of ignoring her messages all day. These forced rows were happening more and more. At the time when I needed her support and counsel more than I ever had previously, I found myself inexplicably trying to cut Mary off. There was a very large part of me – consciously and unconsciously – that wanted to push all the people near to me as far away as possible. I wanted to wallow in a muddy quagmire of my own self-pity. Rather like that feeling when you think everyone's forgotten your birthday: exquisite, self-righteous misery. 'Nobody loves me but I don't care!'

But Mary understood all this. She got it. And she put up with it. She gently coaxed me out of my unwelcome thought patterns when I let her, and more forcefully when I didn't. ('Oh Lord, is she fierce?' my father once asked me.) She instilled in me a belief in my own ability to survive and, most importantly, she gave me hope for the future. She too comes from a family of disparate sparrows, perhaps a little less noisy and flappable than my own, but great flocks of diverse siblings, cousins, uncles and aunts. She'd also had to cope with the death of her father at a young age and had to fight to keep her own immediate family together, and so she became not just a wife but an indispensible best friend and mentor. She is practical where I am quixotic. Unflappable when I flap. But even Mary I strove to keep at a distance.

When I wasn't out-and-out rowing with her, I set myself to 'politely distant' mode and carried on. Because, quite frankly, I believed the only person who could get me out of this mess was me. Why trust anyone else? And it wasn't just Mary. I ran the shutters down on Richard and Katie too. I stopped taking their calls and I very rarely met with friends for drinks either. And so I stumbled on through my angst-filled days, battling the world, pushing friends and family away, and putting myself in a kind of emotional quarantine. And then one day, quite unexpectedly, the walls came crashing down around me.

It was a few weeks after the night of Mum's birthday. Mary was overseas for work and I was dangerously home alone. On the face of it, it had the potential to be rather a cosy life-affirming sort of night and ordinarily I look forward to those evenings immensely. Complete unhindered tranquillity, eating crap food, drinking nice wine and watching the sorts of warm and cuddly nineties romcoms I'd be embarrassed to tell my friends, or even Mary, I'd watched (she has a phobia of Tom Hanks).

And that night started pretty well along those lines. I shoved a frozen pizza in the oven, lit a fire and poured myself a family-sized portion of gin and tonic. Unfortunately for me, though, I never did manage to watch the life-affirming romcom.

Instead I drank steadily through the evening and became progressively melancholy. I was overwhelmed by my thoughts, my doubts and insecurities, endlessly and pointlessly questioning my place in this strange and increasingly unfamiliar world. Everything in my life that had felt so inalienable and certain, it turned out, was just a mirage. All the rock-steady beliefs, by which I had defined my sense of self in days gone by, had crumbled. No matter how depressed I'd felt in the past I'd always felt like my life had some kind of purpose, and I was pretty confident about where I fitted in the grand scheme of things. But that was just a load of old bollocks really. It turned out my existence was without purpose and, worse, this grand scheme everyone talks about endlessly is just a giant hoax. It doesn't exist. It's just endless unplanned space. I convinced myself that there was no one who could help me or ever possibly understand. For some reason this made me feel like a total fraud. An act. But I had to keep people believing I was the person they believed me to be, this character I had created: the cheerful and philosophical man of the world – one of nature's old souls. But it was just a brittle shell.

All these profoundly unhelpful thoughts were ringing around my head like so many giant church bells – clanging so loudly I couldn't cope with the din. It was then, for the very first time in my life, that I began to think seriously about suicide. In the past suicide was only something I had thought about in the abstract: better to jump off a cliff or take an overdose? But this night was the first time I genuinely felt there was no place for me on the earth and all those around me would be far better off if I shuffled off this mortal coil. I sort of collapsed gently into myself. And had I been given the means to do something about those thoughts that night, who knows? I was probably too pissed anyway. Shortly after, for some inexplicable reason, known only to me then, I turned all the furniture upside down.

I began to feel, as I sat amid the upside-down chairs, sipping my gin, that maybe suicide was genetic. After all, there was a very strong likelihood my uncle, Mum's brother Michael, had taken his own life. As much as Mum had always denied even the slightest possibility of that. Is depression in your bones? Was I born with this entrenched sadness or was it possible for me to find a way back up the staircase and into the sunlight? I just did not understand any longer how to be happy or why anything mattered any more. And the strangest thing of all was that I could not link my renewed anxiety, or even Mum's death, with this feeling. I just saw it as my own self-indulgence. My own innate weakness.

I awoke to the soundtrack of my sparrows. God knows what time it was. A time when I should have been on a train to work or perhaps even sitting at my desk sending worthy business-like emails to colleagues. I was fully clothed and spread-eagled on the bed, staring unsteadily for quite some time at a full glass of water on

168

my bedside table. And then, with my brain frozen in aspic, I thought to myself: *It's too light. This is a mid-morning kind of light, not a 6 a.m. get-on-with-the-day kind of light.* That was work out of the window. I looked at my phone; there were a healthy amount of missed calls from Mary. Good. Somebody cares. And five missed calls from Dad. Bad. Can't be dealing with Dad today.

I drank the water by my bed at a rush in a vain attempt to unglue my desiccated tongue, which had somehow overnight become cemented to the roof of my mouth. I crept shakily downstairs, resolved I'd call Mary when I had located my brain and brewed life-saving tea. In doing so I passed the wreckage of my 'quiet night in'. I found I had smoked about 6,500 cigarettes and consumed 3.5 gallons of gin (and a bottle of tonic). Oh God, and a bottle of red wine too! When did I drink that? The room was strewn with the detritus of the lonely drunk. Empty fag packets, overflowing ashtray, discarded bottles and iPod, plus headphones, still playing Peter Sarstedt's 'Where Do You Go To (My Lovely)?' on repeat. And why on earth was all the furniture upside down?

Yet still the sparrows chirruped cheerfully outside. They'd taken control of my bird feeders, battling off any other little bird that dared come close. Cheeky buggers. Impudent even. I watched them for several minutes. Not so much because I was entranced and overwhelmed by their beauty, but because I found it easier to look at them, with my pounding head and aching eyes, than anything else. They didn't require me to think, those sparrows.

They just carried on doing their thing. And I did my thing. I can't honestly recall whether I had any kind of soaring revelation then, watching those bustling, chirruping little creatures. I was too busy focusing on keeping my brain from leaking out of my nose. But they inspired in me the idea that I should go for a walk. Or, at least, I should force myself to go for a walk. My instinct was to lie on the sofa with two bottles of Coca-Cola, feeling sorry for myself until the next sunrise. But I knew that if I did that, I'd quickly be overwhelmed by my thoughts. And that didn't bear thinking about.

My walk took me through the village, down to the canal, and round and back again. It takes about forty minutes and the locals call it 'the circuit'. It's the kind of walk where you always bump into someone you know: a cheery hello here and a scrap of gossip there. Fleeting human contact that warms the soul and, just for a moment, distracts you from your pounding thoughts. Not that I had wanted to see anybody at all that day. Or ever again. But there was a tiny little voice, deep down, that urged me onwards. *Do it for the birds, if nothing else*, it said.

In fact, so familiar was I with this little walk that I knew every bird and exactly where I might see it: a robin as I walked down the track by the church, followed by a wren just on the corner as I turned west towards the canal, usually followed by a couple of blue tits – a great tit or two – and a flourish of blackbirds squawking proprietorially. Next was the bullfinch run. The 100-yard stretch of track that wound up to the canal was prime territory for my bullfinches, and invariably I'd see or hear one piping in the hedge. As I did that day. If I was lucky, I might also see a tree sparrow down the track (the house sparrow's quite shy rural cousin) as well as a chaffinch or two going about their daily chores.

Then when I made it to the canal there was always the heron by the bridge, ducks and moorhens floating about minding their own business, and, if it was a special day, I might be lucky enough to spot a kingfisher. And always, when I got back from my walk, there were the sparrows in the village, criss-crossing the hedgerow on each side of the road as I neared my house. Always there. Always together.

I needed to learn how to live again. Just to get on with my life. Like those sparrows. I simply could not allow myself to get that low again. It was the Night of the Blackest Dog and I really, really could not risk a repeat performance. Outside of easing up on the gin, I realized that I needed to reconnect with the sparrows in my life. I needed to find the courage to open up somehow – to talk about this stuff, without it feeling like an act of great shame or weakness.

And so it started that very day. The process. It is a very long process and I am still working my way through it. But the turning point came that day, on that rural lane between two bare hawthorn hedges and with the sparrows criss-crossing ahead of me. I took the decision to stop running away from my fears. Slowly but surely, and with more than a few abrupt U-turns, I forced myself to go to meetings, to barbershops and even church services. I deliberately began to put myself in positions where I knew a panic attack was inevitable. At first it was horrendous. But it got easier over time. It was baby steps, of course. And it did feel a little bit like the emotional equivalent of those heroic RAF officers I used to see in black-and-white war films that had to learn to walk again after a bad crash. After a time I even summoned up the confidence to get up on a stage, which had always been such an integral part of my old job and something I really never thought I would be able to do again. That was a momentous day.

In parallel with being brave about haircuts and public speaking, I also decided to talk to people. A genuine turning point came many months later when I decided to go back to my American chap in Victoria, and just be honest.

'The problem is, I get these terrible attacks that I can't really explain, when I am in meetings or any place where I can't escape.'

American chap: 'Why?'

Me: 'Erm, well . . . you see. Erm. Well, the thing is. Erm. I think I am going to have an accident and soil my trousers.'

CUE PROFOUND SHAME AND EMBARRASSMENT.

American chap: 'So what?'

Me: 'Eh?'

American chap: 'So what? Has it ever actually happened?'

Me: 'No. But that's not the point. It *could* . . . and it would destroy my career.'

American chap: 'Why? What is the very worst that can happen, Charlie? You shit yourself and they don't ask you back.'

Me: 'Yes, but – Well. What if I fart? Sometimes I have a great fear of just farting.'

American chap: 'Fart then.'

My life was transformed. Mr American told me that what I was suffering from was, in fact, very common and brought on by deep levels of anxiety. He explained the fight or flight instinct that is engrained in the human psyche from the days when we were chased by large fearsome mammals with big teeth. What I was experiencing was a natural part of that.

'When you feel an attack coming on, take two very large breaths. That will help to reduce the fight or flight instinct.'

It all made complete sense to me when he put it like that.

'In fact, the head of one of the biggest multinational companies in the world has been here to see me about it in the past. It used to happen to him in shareholder meetings,' he said.

Well, that made all the difference; I wasn't alone. But, even more reassuringly, I was also in the company of hugely successful chief executives. What had been a fearful and totally inexplicable event in my life, which had terrified me and left me feeling such a great weight of shame, became an ordinary part of the human story. It had a basis in science and reason.

I left his office feeling like a million dollars. Like the

biggest, moon-sized burden had been lifted off my shoulders. Light as a feather. And like a bit of a high flier too.

I started experimenting with telling people. I told Mary, and she didn't laugh or think less of me. I told Richard, and he said it had happened to him too occasionally. I told Katie, and she just hugged me even tighter. I just kept telling people. I told my brother-in-law, Ed, and he understood. In fact, no one seemed to mind or think less of me. I made a joke of it all and told more people. And the more people I told, the less it seemed to matter. And now I'm telling you.

There's a great quote from the film *Crocodile Dundee*, and I have begun to live by it. The reporter who's brought Mick Dundee to New York is telling him about her friend who goes to see a psychiatrist to talk about her problems, and Mick turns to her and says, 'Hasn't she got any mates? Back there if you got a problem you tell Wally, and he tells everyone in town. Brings it out into the open. No more problem.'

I often ponder how easy it is to ignore and neglect the creatures and people closest to us. It's a dangerous game to play. I never used to notice the cheerful chirrups of the house sparrows around me in the days before I made an effort to reconnect. They were just so much background noise: drab little white-and-brown creatures with little to offer in terms of awe or wonder of the natural world – hardly an osprey fishing or a golden eagle in flight, right? How wrong I was. Those sparrows are integral.

As a species, we humans have ignored the everyday, unglamorous nature outside our doors for so long now that's it become like just so much old peeling wallpaper. But our species' dislocation from its local environment, from this everyday nature, has consequences. We are living at a time of unprecedented loss of once abundant, now vanishingly rare, birdlife. In the case of the house sparrow, numbers have fallen in some places by up to 70%. And to be quite frank I'm not entirely sure many people have noticed.

From these sparrows I learned one incredibly important lesson: just to live. To take solace in the everyday. Even in the grey skies of an empty Tuesday afternoon – a prime time for endless thinking and creeping melancholy. But a sparrow doesn't know it's a Tuesday afternoon. Nor does a wood pigeon or a starling. They just get on with what needs getting on with, oblivious to my inner struggles. Or as the philosopher Alan Watts once said:

> The meaning of life is just to be alive. It is so plain and so obvious and so simple. And yet, everybody rushes around in a great panic as if it were necessary to achieve something beyond themselves.

God bless you, Alan. And God bless those humble house sparrows.

Not that long ago the house sparrow – that most impudent, nay saucy, of characters – provided the background music to the life of town and country dweller alike; the cheerful chirruping never ceased. But no more. They have vanished from our streets, and even from large tracts of the countryside, too. The silence is deafening.

And what is so strange is that house sparrows used to be such enduring little blighters. You could never get rid of them, no matter how hard you tried. They were at complete ease living alongside people: nesting in our houses and eating all the veg. I can remember my parents' anguish at the great clouds of sparrows that would erupt from our vegetable garden when we opened the door in the morning and caught them unawares. In fact, such was their resilience, that even a Second World War campaign by the British government to annihilate them had absolutely no effect at all. It was felt by the powers that be that the humble sparrow was eating too much fruit and veg during a time of rationing. People were actually paid per sparrow they killed. Despite the ensuing slaughter, the sparrow population remained as strong as ever.

Just how have my sparrow friends turned from resilient, impudent pest to struggling underdog? It is a question that regularly knocks on the door of my mind as I wander the silent sparrow-free streets and lanes. Some people blame loss of habitat, while others blame overpredation from sparrowhawks and cats. Some people even blame unleaded petrol. Personally I think it is a mixture of the first two theories. (I'm not entirely sure how you prove the last.) Sparrows used to thrive in town and country because in the old days our houses were more tumbledown affairs, with gardens full of grass, flowers and weeds. Sparrows like to nest in cracks in the wall, under broken tiles or tucked into a loose brick, and they eat grasses, grains and seeds. We no longer tolerate this sort of idiosyncrasy in our houses. Our gardens, more often than not, are lifeless low-maintenance lawns. Green deserts. We want to live in secure, hermetically sealed little boxes that are easy and cheap to run. Every crack filled, every tile fixed and every stray brick neatly grouted. No blade of grass out of place. This keeps us warmer, reduces our heating bills and we feel safe. But we've left no room for nature. Put simply: there is nowhere any longer for sparrows to build a nest and nothing for them to eat. We've evicted them from our lifestyles without even realizing that we've done it.

I think that one of the reasons our forebears had a much closer relationship to the birds around them, apart from the fact there were just so many more birds – in variety and number – in those days, was that they collected their eggs. Back in the eighteenth and nineteenth centuries (and probably before) every child in the land educated him or herself by looking for nests around their gardens and neighbourhoods and taking a single egg from each one until they had a complete collection of every local species. It was a sort of nineteenth-century version of Pokémon cards. I am certainly not advocating egg theft as a viable hobby for children today, though. It has been illegal to steal wild birds' eggs for many years. Quite right too.

But all this nest searching and egg snatching by the children of old familiarized them with the wildlife that, at that time, lived in abundance all around them. They got to know the bird and its surroundings in a way that was intimate. Through searching for nests they gained a real insight into the life of the bird and its relationship to the environment in which it lived. They learned what it looked and sounded like, its behaviour, at what time of year it nested, what that nest looked like and, of course, what each different species of birds' eggs looked like. It was a cornucopia of life-affirming knowledge.

Quite often I will contemplate what equivalent I can suggest for my young sons in order to inject the importance of wildlife powerfully into their consciousness – but in a way that is fun for them and doesn't involve breaking the law. I don't even want to encourage them to find nests at all, to be quite frank, let alone steal eggs, because finding nests alerts local predators to their existence. A keen-eyed magpie, crow or crafty cat will raid the nest once it's seen you pointing and staring, or the bird will desert the nest because it doesn't like the idea of you poking about nearby.

All I am left with is my enthusiasm and the hope that it rubs off on the boys. I figure that if I become a kind of incessant background noise, somewhat irritating but always there, one day, when they are adults, they will find they somehow already know that that is a chiffchaff singing in March, or a green woodpecker snaffling ants in the garden, by some form of mental osmosis, and without them having to make the effort to learn.

I want them to know this stuff because I want to equip them with the emotional ballast that an innate knowledge of the ecosystem in which they live can provide. I want them to embrace the perspective that nature gives us. I want them to see themselves as an integral part of the natural story, not just a bystander staring at the pictures. And in the case of these sparrows I don't want them to 'encounter' them 'near human habitation'. I want my children to live and breathe the sparrows, to watch and wonder at them living among us each day.

Sparrows chirruping, wood pigeons cooing, thrushes and blackbirds singing heart-warming solos day in, day out formed a kind of background symphony to my childhood springs and summers. And yet today on the farm on which I grew up only the wood pigeons and blackbirds remain in reasonable strength – the sparrows and thrushes are almost gone. Starlings too have disappeared. When I was a boy the starlings were so numerous they would turn the sky black at certain times of the year; it was magical. Dad would lumber out with his shotgun into the gloaming and bang hopelessly away at them in a bid to control the numbers, but it was comical to see how little damage he could do. And yet today you'd be lucky to see more than ten or twenty starlings in the same place at the same time of year.

But every so often those starlings can still work their magic. Not long ago I rushed into the sitting room, grabbed my five-year-old son (ignoring his strong protests) and dragged him outside.

'Look! Look, Arthur!' I whispered urgently as I pulled him out into the garden to witness one of nature's oldest miracles.

Our beech tree strained under the weight of God knows how many hundreds (thousands?) of starlings. It was a gyrating mass of black and purple – more starling than tree. The haunting chuckling noise of those massed ranks of birds – combined with a lurking mist – gave our garden a mystical feel. And then, all of a sudden, the chackling stopped. For a millisecond the world stood still. Before the tree ahead of us exploded in a thunderclap of wings. And the starlings were gone. Most important to me was that my son saw this wonder of nature. That he felt its power. That he realized there was more to life than Buzz Lightyear and Batman.

Murmurations of starlings are seen less and less these days, but when starlings fly they take the form of giant shoals of fish floating in the air. It was just a fraction of a glint, but as I pointed out those starlings to Arthur I'm sure I saw that little boy's soul twitch, just a bit, before he dashed back, barefooted through the mist, to his Lego.

I realized at the time of Mum's death that nature had become like an old forgotten friend I hadn't seen for decades. We humans, and the wildlife that surround us, have become strangers at the great cocktail party of Life on Earth. And, let's face it, it's much easier and more enticing to watch David Attenborough explore the jungles of Madagascar on the BBC at 7 p.m. on Sunday, on a comfy sofa with a drink in our hand, than to go outside and form an attachment to a local squadron of plain brown-and-white sparrows chittering away on the road. But they are nature's ballast, and once they're gone, we won't get them back.

Like everything in life, once you get to grips with the fact that you matter – actually matter – and whether you like it or not you have an impact on those around you, people *and* nature, then it is much easier to cope when things go wrong – or even if you're just going through a bout of random black despair brought on by you-don't-know-what. I make a point to work much harder at my relationships, both with the sparrows and those dearest to me, because I realize that if I become complacent, I will lose them. And, once I do, they might never come back. One day I'll look out of the window and the sparrows will have disappeared from their dust baths on the road, and the pathway to my door will be empty.

These days I've managed to get my anxiety under control. And I've also worked out that anxiety and depression are inextricably linked. It just never occurred to me to think of it like that before, to connect the two together. Anxiety results in depression. Depression results in anxiety.

The unseen hockey stick of fate still flaws me regularly, smashing my ankles away from behind and sending me flying, but I've learned anxiety and sadness are a part of the natural order of my life. No longer terrifying imposters invading my peace of mind, but an innate part of my make-up. Of who I am. And I talk about this sort of stuff with old friends and family. I keep them much closer to me, these people, if not physically definitely emotionally. They are just a phone call away. I still find it difficult, I'll be totally honest, all this sharing business, what I would have seen not that long ago as self-indulgently 'baring my soul'.

My first instinct is to bury all the pain and put on the mask, like my mother did. *I am a self-sufficient human being,* I will tell myself. *I don't need help. I learned that at prep school.* And at times I still sneer at myself for not being able to solve these problems on my own. But I am wrong. I know that now.

When I force myself to open up to close friends and family, I am never disappointed. And no one has ever yet, that I know of, thought the less of me. And I don't care if they do, to be quite frank. I no longer see it as shameful to admit to crushing anxiety or inexplicable sadness – or even needing to go to the loo in the middle of a meeting. It is what it is. And by taking this line of thought I find it so much easier just to live – like a sparrow you might be lucky enough to encounter dust-bathing near human habitation.

House Sparrow

What it looks like: *A riot of browns, blacks and greys, the house sparrow is a small scruffy-looking bird that hangs out in large mobs. The males have a grey crown, a prominent black eye mask above their stubby beaks, streaky black and chestnut wings and a grey belly. The female sparrow is mainly brown and grey, without the black eye mask.*

What it sounds like: *House sparrows cheerfully chirrup-chirrup-chirrup all day long and through every season.*

Where to find it: *The house sparrow has always liked to live among humans, nesting in holes and roof spaces in and around houses, farms and barns. It can be found in town and country alike, on streets, urban parkland and farmyards. It is not a bird of the wide-open spaces. Although its cousin the tree sparrow (a neater version with a chestnut crown) feeds in large arable fields.*

What it eats: *The house sparrow has a wide diet of seeds, weeds, nuts and berries, as well as insects and kitchen scraps.*

Chance of seeing one: *Up until twenty-five years ago seeing a large gang of house sparrows on any city street, town square, village or farmyard was a 100% certainty. However, in some parts of Britain house sparrow populations have fallen by up to 70%. No* one can agree on the exact reason for this decline but it is thought that a lack of nesting places due to modern building techniques, less food due to pesticides and intensive farming, and overpredation by cats and sparrowhawks play a big part. You've still got about a 70% chance of seeing a sparrow anywhere in lowland Britain, town and country, but the gangs will be far fewer in number than in days of old.

8. Chiffchaff

Spring is still spring. The atom bombs are piling up in the factories, the police are prowling through the cities, the lies are streaming from the loudspeakers, but the earth is still going round the sun, and neither the dictators nor the bureaucrats, deeply as they disapprove of the process, are able to prevent it.

> George Orwell, 'Some Thoughts on the
> Common Toad'

> See at yon flitting bird that flies
> Above the oak tree tops at play,
> Uttering its restless melodies
> Of 'chipichap' throughout the day.
> John Clare, 'The Chiff-Chaff'

The chiffchaff is a nondescript olive-green bundle of feathers, with a very faint yellow stripe above its eye. It's about the size of a squash ball and almost impossible to identify by sight, as it flits delicately, soundlessly through the treetops. It is one of the first warblers to arrive on our shores from Africa in March, and it nests at the foot of tall grasses and bushes. But if you do spot one, it's worth contemplating for a short moment that the chiffchaff weighs less than half an ounce and yet it has flown all the way from West Africa, across the Sahara Desert and the wide Mediterranean Sea, before flitting past you on your sedate woodland walk.

Not much to write home about looks-wise, I'll admit; it's hardly a proud chaffinch with his vibrant copper breast or a dainty blue tit with his exquisite yellow and blue livery. But the importance of the chiffchaff is not in what it wears but in how it sings. And *when* it sings. The chiffchaff's song – its loud and persistently cheerful *chiff-chaff, chiff-chaff, chiff-chaff* – is nature's springtime alarm clock. Once you've heard that first chiffchaff sing out in March, you can be sure that the season is on the turn and that the natural world is about to wake up from its winter slumber and explode into life. In the months after Mum died those happy notes assumed a new significance for me. They broke a silence between my father and me, that had so very much needed to be broken.

Spring came late that year. Winter clung on like an old drunk at pub closing time. It set in and wouldn't be budged. Just when I thought spring was about ready to greet my day with a warm embrace, yawning and stretching, winter would make an ugly face at it and spring would scurry fearfully back beneath the bedclothes.

Not long after my Night of the Blackest Dog I made a trip to Dad's farm to see how he was getting on with the lambing. I arrived one brisk Saturday in mid-March. All around I could sense the sap on the rise. The earth was erupting with profusions of little yellow flowers and underneath the oak at the end of Dad's drive, January and

February's snowdrops had given way to a carpet of golden March daffodils. But there was a bitter chill in the air and between the weak sunlit spells a freezing sleet gnawed at my fingertips. Winter was in no mood to retreat quite yet. *I am the spirit that denies*, as Goethe might have said.

It was the first lambing season without Mum, which was going to make things pretty tough for Dad. I remember it as being one of the hardest, coldest, darkest times of the year on the farm. The generally accepted view of lambing is that it is a miraculous time of year, with little lambs gambolling around sunlit fields, against a backdrop of gently swaying daffodils and rousing birdsong. And there is truth in that. But the picture this pastoral fantasy fails to paint is the bit before the sunlit fields and skipping lambs. The hard bit.

Every year for six weeks or so in my childhood the lambing would overtake our lives. We lived in a constant atmosphere of stress, fatigue and the overpowering stench of new life. The kitchen would take on the feel (and smell) of an agricultural maternity unit, overflowing with bottles, teats and powdered milk, plus mud and sheep shit, as a fairly constant stream of orphaned lambs took up residence in straw-lined boxes next to our warm oil-fired stove. As children it was our job to bottle-feed these starving little bags of flesh and wool, whose mothers had either died or had rejected them. My brother's Jack Russell terrier would gently lick them back to good health or soothe them back into the fold of their creator. It never ceased to amaze me how this loyal little dog, who would rip apart a rabbit or a rat (or a postman's trousers) at the drop of a hat, transformed into caring nursemaid for these pathetic little bundles.

Because Mum had small hands she was designated chief midwife during the lambing season, and because the ewes of the particular breed of sheep Dad kept were known to be 'difficult lambers' it meant she was on constant duty most nights up at the lambing pens. Those March nights could be brutal. Dad certainly did his share at this difficult time but it always felt like Mum took most of the burden. She would be up at the pens at midnight for a final check on the heavily pregnant ewes before returning at 3 a.m. to check for any unwanted developments. And there always seemed to be any number of unwanted developments, generally requiring a hand up the rear of a sheep. In fact, such was Mum's dedication to the cause that she once lost her wedding ring up the back of a sheep.

I really wasn't sure how on earth Dad was going to make it through this particular lambing season without her support. Not just because he'd been unable to fit his hands up the backside of any ewe that was experiencing difficulties with a breached lamb or worse, but because grief had taken a serious toll on him. It had robbed him of his spirit and if he didn't start to eat properly soon, it would rob him of his health too.

Despite Dad's neighbours providing him with a steady stream of cottage pies left daily at his front door, he never touched them. His diet of gin and bananas was really beginning to show. This once portly and energetically red-faced man – who would hoover a full English every morning, a sandwich and a pint at lunch, and meat and two veg with half a bottle of cheap claret for dinner – had become a pale and gaunt reflection of his former self. As he shuffled up the small hill towards the lambing shed, head bowed, he had become barely recognizable. His tatty old boiler suit flapped and slapped round his bony frame in that bitter March wind, and he moved with a kind of broken and stumbling gait, like a lame donkey on the way to the knacker's yard.

It was a tortuous few hours. I just couldn't get any spark of life out of him. He was taciturn and sullen. And, in desperation, I made the schoolboy error of asking him how the lambing was going. Never ask a sheep farmer how the lambing is going, particularly if their wife's just died and they haven't eaten properly for three months.

'Bloody awful!' was his response. And he went on to list a depressing catalogue of misfortune, from prolapses and disease to death and low numbers. 'We've barely had any twins or triplets at all this year!' Farmers never admit to good news or prosperity, only bad news and poverty.

After about half an hour of filling water troughs, spreading straw and lugging hay bales about, we made our way slowly back to the house. Dad suggested we go and visit Mum's grave. I'd not risked visiting Mum's grave since her funeral. I wasn't quite sure how it would take me. Mum is buried in the graveyard of a plain, solidly constructed eighteenth-century church at the top of a hill, just outside the little hamlet where I grew up. It's a lonely spot, with a smattering of ancient yews, two or three towering beech trees and a spreading view across a patchwork of hedge-lined fields. I was christened in that church, and I've yet to find somewhere I'd rather be buried.

The funeral, just two months before, had been a day of great sadness and unexpected cheer, cold reality and warm remembrance, with an old-fashioned funeral in the morning and a service of thanksgiving in the afternoon. We buried Mum at 10 a.m. at this quiet spot with just close family in attendance. The ground was rock hard with a layer of thick frost. I can remember feeling sorry for the people who had had to dig the grave the day before. Odd, the things that pop into your head at these times. Inside the church it was all looming shadows and cold to the touch. The ancient heating system had not yet had time to warm the place, and what little light there was that grey January morning failed to penetrate. It was glum. We huddled together against the cold in the front three pews of the church, doing our best to be stiff-backed and brave – and then failing. We sent Mum on her way in the traditional fashion: *the Lord giveth and the Lord taketh away*. We said our prayers. And then, on an old tape recorder, we played several of Mum's favourite Scottish folk songs. Mum had always been fiercely proud of her Scottish heritage. I closed my eyes and could see so vividly this energetic redheaded woman ('Auburn, darling!') catapulting herself around the dance floor, MacDuff tartan sash trailing in her wake and with great peals of laughter ringing out. Then I wept. They were the first tears I had shed since her death; I tried desperately to hold them back but they shuddered out of me in fits and bursts. Mary gripped my hand like her life depended on it, tears rolling down her cheeks.

But then in the afternoon we held a service of thanksgiving at the parish church in the neighbouring village and our world changed from the sombre black and white of the morning to full, glowing Technicolor. I arrived about thirty minutes before the service was due to begin to find total gridlock in the village. The church simply could not take the weight of people who had turned out for the service. They spilled out of its doors and vestibules, filling the graveyard that encircled the church, and creating a feeling that someone who genuinely mattered – someone who in her own quiet and effortlessly kind way had changed lives for the better – had gone away for good. 'Bloody hell, Mum. This is quite something,' was all I could think of to say.

The service itself passed in a heady blur. A sea of warm bodies and bright colours set against the chill monochrome of the morning. We sang 'Lord of the Dance'. It had been Mum's favourite hymn, and a more fitting tribute to her I cannot find. I have never sung a hymn with more ferocity and meaning than in that church on that day.

'And I came down from heaven
And I danced on the earth.'

I was so incredibly proud that my dear mother's funeral shut down our local village, panicked the local constabulary and caused a three-mile traffic jam on the winding lanes of my childhood. That was Mum's traffic jam. Our traffic jam. ·

And then we drank champagne. Vast, bubbling quantities of the stuff. We did this without guilt or a second thought, because it was the single action that we all knew, with absolute clarity, that Mum would genuinely have wanted. At no point during her short illness did she have any intention of dying, God, no, but she did make one point very clear: if I do, there will be champagne at my funeral. My father took this very seriously. In fact, champagne was the only drink on offer that day. And we all drank it to the end of excess.

And that had been Mum's funeral. Two months felt like two years, as I stood in that graveyard on that chilly March morning. Dad laid some daffodils at the foot of Mum's headstone, which had only just been erected. 'Jolly smart, Dad,' I said. Smiling inside again at the inscription he had chosen: . . . *darling wife of Peter Corbett for nearly forty-four years.* I realized then, of course, that today was my parents' wedding anniversary. If Mum had lasted another two months, there'd have been no need for the 'nearly'.

While I found it totally impossible to have a conversation with my father on any subject, let alone grief or its aftermath, I found deep reassurance in the strength of his love for my mother, his wife.

'There's plenty of room on there for me, too, you know,' he said, pointing to the headstone. 'I made them dig a twelve-foot hole [poor sods]. Make damn sure you bury me here.'

'We will, Dad.'

I can't remember how long Dad and I stood there up on that lonely hill, awkwardly together, with the chill wind whipping at our ankles. I contemplated the last few bitter months and the scars they had left behind. Grief, like winter, leaves an altered landscape behind it for good and for ill. Standing next to me was the shell of the man I once knew, dangerously weakened in spirit and in body. Would I ever have dreamed this possible just a year ago? What had I been doing this time last year? That's right. Mary and I had not long returned from a two-week delayed honeymoon in Mexico. I had never felt so well rested and alive. I was bursting at the seams with happy anticipation. I felt confident, loved and secure. And yet here I was, just twelve months later, standing at the grave of my mother next to an emaciated and spiritless version of my father. It would have been an unimaginable development to that cheerful and optimistic newlywed man of one year before.

I looked across at Dad and I thought about how many memories this little plot of land must now hold for him. His children baptised here, Katie married, his wife buried here, and, oh dear God, the remembrance fell on me like a heavy stone slab: his daughter Emma.

No one ever mentioned Emma. She would have been my sister. She died before I was born. A cot death. It had been another closed chapter in Mum's life. Another deep torment and tragedy, cruelly mounted upon the death of her brother and father, that she never spoke of. And neither had Dad. I had always thought of it as something Mum went through, but how had it affected my father? Did he feel the same kind of pain? I will never know. I have never asked. Back when Emma died, aged fifteen weeks, the powers that be refused to let her be buried in the traditional fashion. 'That doctor was a bastard,' my father said when I asked him about it while writing this chapter. And then he very quickly changed the subject.

Mum had wanted to bury Emma here up on this hill, but they wouldn't let her, so the council had taken her little body away in a van and incinerated it. And then returned the empty carrycot. The bastards. Instead of a grave all my parents had to remember her by was a small plaque inside that solidly constructed church, just by the door, on an inconspicuous collection box cemented into the wall. All through my childhood I didn't even know it was there. There were just so many of our memories carefully stored in this small piece of Hampshire ground.

And one day, I reflected, many years from now, when all of us are long dead, people will visit this church and they will see the humble bronze plaque in memory of the sister I never met and wonder who was this little girl? And they'll think how awful it was that she died so young. How awful it must have been for her grieving parents. Like I do now, when I see those ancient graves inscribed with long names, which sound so strong and permanent, but belie the tiny lifespans inscribed below them.

I cannot possibly even get close to pain on that level. I cannot get close to what my mother must have gone through. How it must have changed her fundamentally. How does anyone get over an experience like that? And because she never spoke of it I will never know. But one thing I do know is that somehow she overcame it. And from that knowledge I draw great strength. One way or another Mum, and Dad, found a way through the darkness, and she and my father had another child, me. And no matter how big a mountain it must have felt like at the time, they scaled it and moved on with their lives.

I've very often contemplated that if Emma had survived, I would never have been born. Her death led, two years later, to my life. And while Mum was a mass of contradictions – the warm smiles that hid the cold, unremitting pain – and I had no idea truly what she felt inside, I can be absolutely sure that shortly before the cancer gripped her she was as happy as I've ever known her to be. She and my father had achieved a level of contentment they never had previously. That's the damn tragedy at the heart of the whole thing. Like breaking your ankle just when you thought the race was won.

I stood on that hill and wished so hard that I could talk to my father about these sorts of things. About Mum and Emma. Katie and Richard. And my own sadness. But we didn't. We couldn't. So instead we stood, together yet alone, in an awkwardness of silence. I was about to come up with some weak excuse about needing to get back for lunch, when my thoughts were interrupted by the bright clear notes of a March chiffchaff.

Chiff-chaff, chiff-chaff, chiff-chaff!

'A chiffchaff, Dad! A chiffchaff! That's the first I've heard this year!'

He looked up, looked around, and with perhaps the merest suggestion of an inward smile breaking through his wintry facade, said, 'That would have made Mum very happy.' And it would have done, too.

Mum was a creature of the spring. She came alive when the trees began to blossom and the crocuses and snowdrops fought their way out of the rigid soil. If I had to describe her life in a colour, it would be yellow. A rich, golden yellow. The arrival of the daffodils in March was always a very special time. And out would come the camera, year after daffodil-blessed year. I've lost count of how many photos there must be of all us children at various stages of life, and countless dogs through the ages, sitting among the daffodils in the garden. As that chiffchaff sang, though we were silent, I knew that Dad and I were thinking the same thought: Mum and her daffodils. I knew just how excited she would be that this chiffchaff was heralding the turn of the season. And I also knew she'd be bloody relieved that the lambing was nearly over, too. Thank God.

A hefty weight of that heavy March burden was lifted off our shoulders by those joyous notes and, while I had always been deeply cynical of those people who believe their loved ones come back as a robin in the garden shed or a soaring buzzard over an old oak, on that day I chose to believe it was Mum's spirit in the voice of that happy little olive-green warbler. It gave us both a thin shaft of hope, a lift, on that cold early-spring morning. Like my skylark had done way back in August or the bold thrush in the depths of January. Nature had once again reminded me that life goes on, whether we like it or not.

The chiffchaff is the true herald of spring, not the cuckoo, which comes a month later in April. It acts like a tincture of warm June sunshine into my soul, and I happily contemplate how this heavenly sound has been a balm for troubled souls for I don't know how many thousands of years.

The chiffchaff doesn't sing so much as proclaim its presence. 'Hear ye, hear ye. It's a bloody freezing March that feels like it will never thaw, but I'm here to tell you, Charlie, that all will be well!' That's what it said to Dad and me as we stood in the graveyard thinking about Mum. My fingers trembled with the ecstasy of it. As Winston Churchill might have said, it was not the end of my grief. It was not even the beginning of the end. But it was, perhaps, the end of the beginning.

In the first spring after Mum's death I came to regard the chiffchaff as nature's warm-up act before the smash-hit spring show started in earnest. I closed the curtains on winter one night and the next morning the landscape was ablaze with a thousand different shades of green. My spirits, like the sap all around me, began to rise. This was Mum's season. While it made me feel her absence that much more keenly, I also took great comfort in the explosion of life.

Following close on the heels of that early chiffchaff came all the other sweet-singing warblers, making their way back from Africa to the northern hemisphere to breed. And as late March seeped into April, after so many months of silence, the hedgerows and riverbanks of my home county came alive with song, and willow warblers, garden warblers, reed and sedge warblers – to name but a few – took up their summer residences across the land.

My favourite of all the warblers, though, is the black-cap. And, once I had heard my first chiffchaff, in the days that followed I listened intently for the melting melody of the first blackcap. You won't find the blackcap celebrated in much romantic poetry or literature. Poets tend to focus on its more glamorous cousin, the nightingale. I think this is unfair and somewhat bizarre. Because the song of the blackcap is up there with the nightingale, the thrush and the blackbird. He is the Nick Drake of the songbird world; unappreciated in his time. You don't need to take my word for it. Theodore Roosevelt wrote in his autobiography about a woodland ramble he went on one day in the New Forest in Hampshire:

> The most musical singer we heard was the blackcap warbler. To my ear its song seemed more musical than that of the nightingale. It was astonishingly powerful for so small a bird; in volume and continuity it does not come up to the songs of thrushes, and certain other birds, but in quality, as an isolated piece of melody, it can hardly be surpassed.

I look back at that spring and remember it like nature's three-act play. I moved from Act One on that grey March day with the early chiffchaff – the herald – into Act Two, with the ever-growing cacophony of warblers arriving day after day to fill the hedgerows – growing colour and noise – and reaching the rousing Final Act with all the actors performing for me the dawn chorus. The dawn chorus is a genuine miracle of nature. And the most wonderful thing about it is that even those people who know nothing about the birds, or indeed have no intention of finding out, cannot ignore it. It imposes itself on the human race, forcefully and jubilantly, whether we want it or not. It is an unadulterated expression of life on earth at full volume.

For some, being awoken at four or five in the morning by that rising crescendo of song is a source of great misery, as they struggle to maximize sleep for the long day ahead. I will readily admit that at one point in my life I was that person. I've always been a bit of an insomniac. And I'd very often lie in bed at 3 a.m. fearing the dawn. Because I knew that when those birds started to sing all hope of sleep was lost. But as my love and knowledge of the birds around me grew, I began to see how completely irrational my attitude was. You cannot tell yourself to sleep and expect to sleep. Humans aren't built like that. No matter how many blackout blinds and earplugs you buy, you cannot stop the sun from rising and you cannot prevent the birds from singing.

And so I learned to embrace the birds at 3 a.m. If I'm not going to sleep anyway, why not wallow in the glorious song? Because it is truly one of the most miraculous performances nature can bestow, an awe-inspiring show. When I cannot sleep on a spring morning these days, I lie in my bed listening in wonder, and imagine that right at this very moment layer upon layer of birdsong is reverberating joyfully across the whole of Europe, and then I fall asleep.

I had my first genuinely moving encounter with the dawn chorus when I was about eight or nine. I was away at boarding school and housed in a fairly sombre little dormitory that faced on to a small wood. I was roused from my slumber at some ungodly hour by the most deafening racket. I simply could not comprehend it. It felt like a wall of noise. I crept over to the small Victorian casement window and perched on the sill, with not another soul stirring, and listened in shocked wonderment. That was over thirty years ago and I can still hear those songs resonate down the decades at me.

As descriptions of the dawn chorus go, though, it is hard to beat this one from *The Charm of Birds* by Edward Grey. It was actually written by his wife:

> It opens with a few muted notes of thrush song. This sets the tits waking; they have no half tones. There are the sawing notes, the bell notes, the teasing notes, and the festoon of small utterance that belongs especially to the Blue. But you can hardly pick out the individual songs before the whole garden is ringing. There is the loud beauty of the thrushes. Seemingly further away, and in remoter beauty, comes floating the blackbird's voice. The notes are warm, and light and amber, among the sharper flood of song. The dawn chorus is like tapestry translated into sound. The mistle-thrush, with merle and mavis, perhaps the rounded note of an owl, these stand out chief figures in the design. All the others make the dense background of massed stitches; except the wren; he, with resounding scatter of notes, dominates the throng. Then, as suddenly it arose, 'this palace of sound that was reared,' begins to subside.

I read that, listened to the birds and I no longer feared the dawn.

But even that elegant, quite moving description can never really do the dawn chorus justice. It needs to be heard to be believed. It's the same with all birdsong. Writers, poets and composers have spent centuries trying to recreate it in prose, poems and song and, to my mind, never quite achieved their aims. It's not that the prose, poetry and music that birdsong has inspired is not in itself beautiful, it's just that it can never quite capture the feeling I have, deep inside, when I hear a willow warbler sing on a sunny June day or a cuckoo call in April – or that chiffchaff in March.

You cannot truly mimic the beauty of a warbler singing in spring – or how it makes you feel when it sings – be it a garden warbler, blackcap warbler, reed warbler or my dear chiffchaff. I learned that it's the same with grief. Everyone's experience of it is different. Unique. And while there are similar notes for all those who go through a personal tragedy, the order in which they are arranged for each of us is infinite. It cannot be neatly categorized or recreated in any way. Nobody could truly recapture the feeling I had when I stood next to Mum's grave with my dad and heard the first chiffchaff of spring.

It marked an integral part of the journey. A turning point: for me, for my dad, and for the farm. Dad got through the lambing that year by hook and by shepherd's crook, and with the support of his family and his friends. The lambs made it out on to those sunlit fields and the daffodils swayed and the warblers sang, as they always had done, year upon year. Dad and I were still not fixed, not by any means. None of us were – Katie and Richard had their own battles to fight, too – but we'd moved through to the next stage. We were still here and we could consider the possibility that life would go on without our linchpin. She was away now. No doubt dancing a Scottish reel high up in the sky to the sound of a chorus of chiffchaffs.

Chiffchaff

What it looks like: *A tiny olive-green and faintly yellow warbler that arrives in Europe from North and West Africa every March, and returns in October – although many remain in Europe for the winter months. It is not to be confused with the other warblers that arrive in spring, in particular the willow warbler, wood warbler and the garden warbler, which all look very similar if a little larger. The main identifying feature of all the migrant warblers is not their somewhat dull appearance but their delightful song.*

What it sounds like: *The chiffchaff is the first of the warblers to reach Europe and its bright onomatopoeic song is the herald of spring. Its characteristic and persistent chiff-chaff, chiff-chaff, chiff-chaff stands out in March and can be easily differentiated from other birds.*

Where to find it: *The chiffchaff is abundant across Britain. It is a bird of the countryside rather than the town and can be found down most lanes and byways of rural areas. It nests at the foot of tall grasses and bushes, but tends to feed and fly at the tops of the trees with an energetic flitting kind of motion.*

What it eats: *The chiffchaff feeds largely on insects: flies, gnats, midges and caterpillars.*

Chance of seeing one: *If you are walking down any country lane in Britain in spring or autumn, there is an 80% chance you will hear the song of the chiffchaff. Like most birds, the chiffchaff falls silent in July and August.*

9. House Martin

I, a stranger and afraid
In a world I never made.
                A. E. Housman

This guest of summer,
The temple-haunting martlet, does approve
By his loved mansionry, that the heaven's breath
Smells wooingly here; no jutty, frieze,
Buttress, nor coign of vantage, but this bird
Hath made his pendent bed and procreant cradle;
Where they most breed and haunt, I have observed,
The air is delicate.
                William Shakespeare, *Macbeth*

What I love the most about migratory house martins is
that, despite mankind's ability to put men on the moon,
robots on Mars and create intelligent phones that know
what you want for lunch next Thursday, no one has yet
to work out exactly where house martins go each winter.
In fact, in Europe in years gone by, people thought that
swallows and house martins hibernated in the mud at
the bottom of ponds over the winter months. While in
China it was generally agreed that swallows transformed
into mussels in winter. In Russia, the first returning

swallow is greeted with a special song and, the world over, it is accepted that a house martin or a swallow nesting in or near your home is extremely good luck.

The house martin is one of nature's most ingenious constructions. Sleek, compact and devastatingly elegant in its blue-black and pristine white livery. It combines beauty with efficiency, speed with elegance. If Steve Jobs had designed birds, rather than computers, it's the sort of creature he would have built – a design classic. Often confused with their close cousin the swallow (on which more later) the house martin is the smallest of these summer tourists, with a stubby little forked tail and white undercarriage. It won't win any singing competitions, with its cheerful little clickety bleat, but if there were prizes for joyous swooping and gliding and mid-air acrobatics, the house martin's trophy cabinet would be creaking at the sides.

House martins, alongside swallows and swifts, arrive from their wintering grounds in Africa in our northern-hemisphere gardens in April and May. They spend between six and seven months here, breeding and bringing joy, before heading back the 4,000 miles or so to, well, nobody really knows. Where exactly it is in Africa that house martins return to each winter remains shrouded in mystery.

I extract great solace from this shroud of migratory mystery. It intrigues and astonishes me in equal measure and it has cemented the house martin and the swallow into the DNA of human folklore. The sadness I feel at the disappearance of these animals in late autumn is as profound as the unfettered delight I experience on their return in April and May. And let's not forget the swift too, that elegant airborne sickle, which arrives around the same time as the swallows and house martins but returns to Africa somewhat earlier in July. People use the word 'miracle' far too often these days; it's become rather overused, hackneyed even. But the return of the house martins, and their swallow and swift cousins, each spring is nothing short of a bona fide money-back-guaranteed miracle.

The arrival of these African migrants is an event that has been marked by people for century heaped upon century. It is an annual milestone that injects summer powerfully into our collective consciousness. It meant more to our ancestors, of course, because winter in those days brought only cold, hunger and a good deal of lingering death. Seeing that first swallow, swift or house martin in spring had tangible meaning for people then; it meant food, warmth and good cheer were just round the corner.

There is something about this succinct diary entry from the eighteenth-century parson and diarist James Woodforde, dated 18 April 1785, which moves me: *Saw the first Swallow this Season this Morning*. I love the fact that this man, writing over 200 years ago, was looking out for that swallow and felt the urge to mark its arrival in his diary. And that centuries later I can pick up an old book that's been sitting in a wobbly old bookcase for heaven knows how many years, open it up and read those eight words, which leap from the page and connect me directly to this long-dead vicar. I love that we share that moment.

By April, and the arrival of these miraculous little

creatures, I had been through some of my lowest moments – the Night of the Blackest Dog being the most obvious manifestation of my grief – and while I could feel my life was most assuredly moving on, I still had within me this indefinable day-to-day sadness that I just could not shake. An embedded melancholy. The question that dominated my mind was, just how can life ever return to normal again?

When I was young it felt like I had a built-in confidence, deep down, that my life was on a steady, if a little volatile, upward trajectory. It felt like no matter how bad things got, however dreadful, depressed or awful I felt, there was always this tiny little voice inside that told me: *You're young; it's OK. Life will get better. It must. It will. After all, you've got years and years ahead of you to get things right. To get happy.* School, university, first job, marriage – chapter by linear chapter; it had all been in the manual of life I'd been reared on since childhood. Step 1, Step 2, Step 3, Step 4, then: 'Congratulations, Charlie, you're a grown-up now and life can begin. Have a cigar.' But then one day I hit thirty-something (or other) and the years ahead – which had seemed like this everlasting upward staircase – somehow felt cruelly compressed. People I loved started to die. There were no longer any obvious, clearly marked steps for me to take, because some fiend had ripped out the last page of the manual.

This deterioration of confidence and self-assurance went against everything I assumed would be the case when I was younger. It was a disconcerting realization that, in actual fact, every year that went by I felt a little less wise, a little less certain, a little less capable of coping with what turned out to be a cruel and capricious world that, quite frankly, didn't give much of a damn about me or the people I loved. It was hugely frustrating. I felt quite close to A. E. Housman's *I, a stranger and afraid, in a world I never made.* I couldn't put my finger on it.

If it were a month, I could only describe the feeling as August: the suffocating and stagnant dog days of high summer. The fresh wonder of spring is far behind you, the birds are silent, the trees are wilting in the heat, and a kind of indefinable melancholic gloom hangs over the place like a pall. *I have of late – but wherefore I know not – lost all my mirth.* These lines of Shakespeare kept popping into my head (and still do, to be quite frank, with particularly bad hangovers). Back then, though, it really did feel like the world, in Hamlet's words, *appears no other thing to me than a foul and pestilent congregation of vapours.* OK, so possibly not that strong and a bit less early seventeenth century but near as dammit to how I felt in those months after Mum died. I had lost my mirth.

I also lost my job. Well, more accurately speaking, my job lost me. They always tell you not to make any big decisions in the year after the death of a loved one. So I decided to make a massive one by chucking in my flashy job – what had been my dream job – as a special reports editor at a newspaper in London.

Looking back, I cannot fathom why I marched into my editor's office and handed in my cards. It might have had something to do with the fact that after eighteen months or so of working with her she still didn't know my name or quite what I did for the paper. No, it can't have been that. I was bloody exhausted – physically and emotionally – and it felt like the right thing to do at the time. I felt exultant on leaving her office, having just chucked away a job that I knew many people would kill for. Although I was somewhat put out at the nonchalant way she accepted my resignation.

'Excuse me, erm, Charlie (?),' she called after me as I left her lair. 'Could you find someone else to do your job?'

And that was that. I suspect she might have had to ask around a bit to find exactly what job it was that I did for her, but hey ho and on we go, as my mother would have said. I was exhausted, utterly confused about my place in the world, or the point of anything really, and, the icing on the cake, I was without a job or the prospect of one. Not to mention the newly acquired but vigorous attachment to gin and tonics I seemed to have picked up.

One day I was walking alone, quite moodily, along the towpath of our local canal – I think I might even have just had a row with Mary about my precarious career situation and distinct lack of income – when I was stopped in my tracks by the most phenomenal bombardment of house martins. I'd already seen one or two house martins that year, arriving each week in dribs and drabs, but I'd yet to see a large gang of them. I'd yet to experience that great relief that any watcher for arriving house martins or swallows feels when he or she sees a good, healthy and sustainable dollop of them blacken the skies. But that day I did. And what a show they put on for me. I sat on the bank of the canal and watched as they wheeled and dived, forming a holding pattern in the air, and then each one would dive down in turn, skimming the top of the water, taking flies. I sat on the towpath admiring this five-star theatre performance for some twenty or thirty minutes; I became completely immersed. And then I realized that I was smiling so hard my face ached.

Watching those house martins performing their exquisite joy-flight released in me a kind of ethereal calm. The difference between watching these heavenly creatures and, say, my experience of listening to the skylark back in August was that it dawned on me that here was something I had actually been looking forward to incredibly hard. I'd been waiting for this moment for months without ever really realizing it. And I thought about all the thousands upon thousands of miles these little birds had travelled just to be here on this day, at this hour, with me. This feeling hadn't been prompted by anything material – like a birthday or a holiday or a night out with friends (events that no longer gave me any kind of happy anticipation). It was free. It was deeply satisfying. And it was all happening here by this canal, just a few hundred yards from where I lived. My life's quotient of happiness, it turned out, had not come to an abrupt end, and even though I could no longer clearly make out the steps ahead of me, and the book of instructions had been long ago lost, I knew then that my journey would continue. And that there would be new destinations equally as exciting and adventure-filled as the landmarks I had already passed.

As that spring progressed, I watched and followed those house martins with intense interest. I watched them build their miraculous muddy nests – glued to the side of the wall of my local pub, the Golden Swan, getting larger and larger as each day progressed. (It takes about ten days for a house martin to build its nest, though they prefer to reuse old nests from the year before – if the impudent house sparrows haven't pinched them.)

I meditated on how far they'd come and how it is they know to come to exactly where they were born year after year. My car's satnav can't even get me to Woking without at least three 'please make a U-turns' so how it is these animals can navigate to this precise spot from quite so far away every year like clockwork and have the energy to do it with barely a single pit stop, utterly defeats me. And in a hundred years, long after I've gone and my life and problems have evaporated into the cosmos, they'll still come. Some unknowable switch will go off in those far-away house martins, somewhere over the Congo Basin, and the journey to the Golden Swan pub will begin again anew.

These days I am childlike in my adoration of house martins, swifts and swallows. 'Look, look! Did you see it? Darling! Did you see it?' I will shout to Mary as we walk down the lane on a brisk April afternoon, squally showers battering our faces. She is used to this childlike enthusiasm, of course, and will usually humour me.

'Yes, darling, I saw the swallow.'

This is when I become a bore. 'Ooh, well, it's 13 April and I've seen my first swallow. I think that is at least a week earlier than last year. Marvellous!'

House martins, swallows and swifts were originally animals of the cliffs, but over however many thousands of years of mixing with settled humans, they've chosen instead to live and breed among us, in our houses, schools, churches and barns.

Well, they'll live among us if we'll let them. I saw recently that some brute near to me had put wire mesh over the eaves on his barn to prevent the house martins nesting. That's like cutting down your flower border to prevent the butterflies. Nonsensical barbarism.

House martins build their muddy nests on the outside of buildings, under the eaves, or a bit of sloping roof. These nests are miraculous little constructions that hang with no visible means of support on the side of the building. If you're really lucky, you'll get four or five nests in a row, with ever-increasing amounts of house martins (they tend to rear two broods a year) swooping and wheeling round your home all the summer long.

Swallows, on the other hand, build their open nests on ledges inside our sheds and barns, on old bits of furniture or even an upturned bucket. I've known people to smash windows in their outbuildings just to provide ease of access to swallows nesting. These sorts of people are my friends. Swifts prefer a higher vantage point, like the tower of a church or, where I used to live in London, in the old bell tower of the local primary school. I used to count them in every spring. I haven't yet mentioned the sand martin, because they are altogether less common. If you're very lucky and near a river, you might spot a colony of sand martins. They live in sandy holes on the banks of rivers and swarm just inches above the water, nabbing insects. Similar to a house martin in shape and size but less glamorous in their apparel, they are of a dull buff complexion – equally beautiful, though, to my mind.

Before I embarked on my journey of bird rediscovery I used to struggle to tell all these birds apart. As do many others. A common conversation I often overhear – while sitting in the garden of my local pub in the summer – goes something like this:

'Is that a house martin or a swallow?' a woman in pink Hunter wellington boots will ask her husband.

'It's a swallow,' he will reply confidently, his Schoffel gilet quivering gently in the breeze. 'Or maybe it's a swift? To be honest, I'm not really sure.'

The best way to tell them apart is to first remember that swifts are entirely unrelated to house martins and swallows. They're much bigger birds with a long-winged, whirling flight (as I once read it described and can't improve upon) and have stubby tails with a pronounced little fork. Or, put more poetically, swifts are airborne sickles that scythe the air.

Swifts also make a not wholly pleasant screeching noise when they fly – like rusty brakes on an aged bicycle. This led them, in more ancient times, to earn a host of glorious names from our imaginative ancestors, including: screech martin, screamer, squeaker, skeer and skir-devil. Though once you are used to this sound, once you are tuned in to it, then it is incredibly evocative. It's a noise that quite literally screams warm sunny July evenings.

There is also a variety of swift called the white-throated needletail. You're very unlikely to see one in Europe, but I bring it up because it is, without doubt, the best name for a bird I have ever heard. Apart from, perhaps, the butcher bird (or red-backed shrike), which impales its bug victims on a larder of thorns.

Another remarkable facet of the swift is that it spends almost its entire life on the wing, only ever landing to nest. And if they do by some wicked twist of fate land on the ground, they can't take off again. In order to take off swifts need to shuffle to the edge of a high ledge and launch themselves, like some kind of airborne ocean liner. I read somewhere that a swift may fly up to 300,000 miles over two summers – between fledging and finding its next nest site. Just think about that for a bit; it'll hurt your head.

Swallows and house martins are smaller than swifts, but incredibly hard to tell apart when in flight. The thing to watch out for is the house martin's bright white bum, as it darts and swoops above you. House martins are made up of just two colours – blue-black and white – while swallows are blue and white with little red faces. Swallows also have longer tails than house martins and are slightly larger. In terms of telling them apart in flight the only guidance I can really give is that swallows tend to fly lower than house martins and have a floppier sort of flap to their flight. Though, to be quite frank, when you see them all swarming together on a warm late-summer's evening, who cares which is which? Just sit back, relax and enjoy the show.

I could easily have chosen a swallow or even a swift as one of my twelve life-saving birds. Each one, in its own way, has brought delight and solace to me at different times of my life. The swallows of my childhood weighing down the telegraph wires outside my bedroom window in late August, as the summer holidays drew to a close, or the massed ranks of swifts that kept me company when tired and somewhat afraid a long way from home in Lagos, Nigeria. Those Lagos swifts have a special meaning for me, because it was in Lagos, of course, on that very trip, that I had first met Mary.

I've been incredibly lucky in that I've managed, purely by accident, to find myself in two of the major global ports of call for swifts while they were in the process of migrating – on one occasion when they were heading north and one when they were heading south. The first, as mentioned, was in Lagos, a place I had vehemently not wanted to visit but had been dispatched to by a heartless editor to write a report about the Nigerian economy. Though, of course, I now have to thank him. If it wasn't for that rather grey, stooping man, and his determination that I should spend two weeks in one of the world's most dangerous cities, I'd never have met Mary and my life would be unutterably different. I vividly remember drinking my ice-cold Star lager on that hot Nigerian March day, chatting up Mary and being spellbound by the swifts massing outside the window of the Eko Hotel's Sky Bar, thirty-three floors up. It was like bumping into familiar old friends at a party full of strangers.

Years later, quite by accident, I bumped into the swifts

again, down in Cap Ferrat in the south of France, on their way south. Again, not a place I would naturally gravitate towards, with its polished pavements and gold-plated street signs, but I was in for a happy surprise. We'd been invited along to an old friend's fortieth birthday party in late July. It was the evening we arrived that I noticed the swifts. Well, you couldn't really fail to notice them. There were thousands of them: screeling, screeching and diving all around us – a frantic mid-air mob, swarming in the sky, as they readied themselves to make the journey across the water back to Africa. I was agog. Though the saddest thing about it all was that while I was craning my neck and rushing about jubilantly, nobody else in my party seemed to notice. They were oblivious. 'Look, look,' I cried exultantly, 'the swifts! What a magnificent sight.'

'They're quite noisy, aren't they?' was about the only response.

One of the things that saddens me the most about the technologically advanced, supposedly enlightened, age we find ourselves in, is the almost universal ignorance of these daily little miracles and mysteries that happen around us throughout the year. There was a time not that long ago when I was oblivious to it all, too, and I look back at my past self and I feel pity. He was missing out on an entire other dimension. We spend so much of our lives searching for instant gratification, largely via our digital devices, but I've learned that true gratification cannot be instant. It disappears in the blink of an app. I do my best these days to get my kicks from the world that exists away from my screen. Like the arrival of the house martins and swallows, and those dear departing swifts in Cap Ferrat.

When I looked up in the sky and saw those swifts all those years ago, I didn't just see some noisy birds; I saw an impossible 4,000-mile journey. I saw a thousand individual miracles soaring above my head. I tried to imagine where they were going, where they had been and how many would make it back to these shores next summer. It grounded me in place and in season, and I can't tell you how reassuring a feeling that is. It was nearing the end of summer and the swifts were returning. And the noise they made – far from being an annoying distraction, those screams and screels transported me to long-forgotten places in my past. And, in a more prescient way, they gave me pause to anticipate my own future adventures – my own life's odyssey.

I've learned that it's all too easy to fall into the trap of living a one- or two-dimensional life. We look but we don't see. We hear but we don't listen. Nature is the third dimension we too readily take for granted, too readily ignore. I feel so lucky today that I took the time to reconnect with it. Every day, every month, every season, nature has a story to tell me. I tuned in to that story and I've never looked back.

I learned a new word the other day: *yūgen*. And for me it sums up extremely neatly how I feel about the house martins in my life, and those swifts and swallows too. It doesn't look like much, just a couple of quite ugly syllables but it is filled, *filled*, with meaning. *Yūgen* is an important concept in traditional Japanese aesthetics or, in plain English, in beauty and art. It is said to mean 'a profound, mysterious sense of the beauty of the universe . . . and the sad beauty of human suffering'. *Yūgen* suggests that which is beyond what can be said. And I really cannot say it better than that. When I watch the house martins arrive in spring, or leave in autumn, or when I think of my dear mother and the joy she brought in her too brief life, or my own unchartered and at times bewildering journey, though I cannot find the words that exactly define the feeling I have, I now know that I don't need to; *yūgen* does that.

I've since moved away from the village in which the house martins gave me such comfort, but not far away. I can always visit them, and do regularly since it means going to the pub on sunny summer days. Although within a year of moving to my new home, a family of house martins did us the honour of building a nest in the eaves of our house, so we only need to open our back door to experience a good dose of *yūgen*. In fact, we also have swallows and swifts in our new village. I cannot tell you what a glorious feeling that is – the relief and inner peace their arrival inspires in me each year, which gives rise, once again, to some sound ancient philosophy from William Shakespeare, from his play *Richard III*: *True hope is swift, and flies with swallow's wings.*

House Martin

What it looks like: *A compact and elegant summer visitor with a blue-black back, white rump and short, wedge-shaped tail. It arrives in Europe from its African wintering grounds in April and May to breed at around the same time as the swallows with which it is often confused. The best way to tell them apart is by their tails: swallows have much longer forked tails than the house martin's wedge. House martins are also slightly smaller than swallows and have a white rump.*

What it sounds like: *A cheerful series of twittering chirps, often delivered while sitting on a telegraph wire near their nests.*

Where to find it: *House martins arrive in Europe in their hundreds and thousands every spring. As the name suggests, they nest on the sides of houses, building spherical nests made from mud that they glue to the side of walls, under the eaves, with their saliva. Very often house martins will be evicted from their nests by jealous sparrows. Most villages and towns in Europe will have a colony of house martins.*

What it eats: *House martins largely feed on the wing, eating flying insects – in particular flies and aphids. They also like to hunt over water, diving down at low level to catch flies and insects hovering on the surface.*

Chance of seeing one: *Numbers have declined quite startlingly in recent years, particularly in the UK, but you will have a very high chance of seeing a house martin in any village or town in Europe from about May until October. The best time to watch out for them is September, when they mass together in preparation for the long journey back to Africa.*

10. Kingfisher

Then we sit on Cowslip-banks, hear the birds sing, and possesse ourselves in as much quietnesse as these silent silver streams, which we now see glide so quietly by us.

Isaak Walton, *The Compleat Angler*

Nay, lovely Bird, thou art not vain;
Thou hast no proud, ambitious mind;
I also love a quiet place
That's green, away from all mankind;
A lonely pool, and let a tree
Sigh with her bosom over me.

William Henry Davies,
'The Kingfisher'

I think I must have read Kenneth Grahame's *The Wind in the Willows* about fifteen times, mostly as a child but two or three times as an adult too. I feel the need to recharge myself on it every now and again. In fact, I am reading it to my elder son at bedtime right now. For some reason Mole's journey up and out into the world and the cast of characters he meets had the most extraordinary resonance with me as a little boy; I was entranced. The image of Mole *scrabbling and scrooging and scratching* his way up and out of *the seclusion of his cellarage* into the *warm grass of a great meadow* where *the carol of happy birds fell on his dulled hearing almost like a shout* hit me square in the eyes when I first read it. And it has never left me.

Maybe I was for so long that lonely mole looking for his Ratty, waiting for the time in my life when I could truly understand the power of nature to heal and for those words of Ratty about his life on the river to have true meaning for me. *It's brother and sister to me, and aunts, and company, and food and drink . . . It's my world, and I don't want any other. What it hasn't got it is not worth having, and what it doesn't know is not worth knowing.*

When I wasn't reading *The Wind in the Willows* as a child, I'd spend a good deal of time knocking about my parents' house on my own, bored stiff. There never seemed to be anybody about. Dad was outside on the farm somewhere, doing jobs; Mum riding; and Katie and Richard away at boarding school. By the time I was six or seven I knew every inch of that house. I'd explored every dust-filled nook and hidden cranny. There was not a room, drawer or cupboard that had not been thoroughly excavated by my inquisitive little hands. If I close my eyes now and transport myself back to the early eighties, I can still identify each room, drawer and cupboard by its unique smell.

I was a living spectre, ever-present, haunting the place. A part of the house's fabric. It held few secrets from me. Apart, of course, from Mum's locked box, which I would daily shake and rattle, like a wrapped present at Christmas, in the forlorn hope I might somehow be able to unravel the enigma. But outside of the thrilling mystery of that box, it was often the most mundane of objects that would fire my imagination on my lonely excavations around the house. I can remember a set of small china cottages arranged on the shelves of an ancient Welsh dresser in our dining room. I used to take them all down when no one was looking, reverently, and play with them in my bedroom. One day I broke one. And to this day I've never been allowed to forget it. 'They're worth quite a bit of money those funny little cottages,' Dad will say. 'Except for the one Charlie smashed.'

My father's desk was a particular highlight. It was full to bursting with exciting objects, a treasure trove of pointless ephemera: impossibly sharp penknives, old coins with King George on the back, strange keys for long ago rusted locks, tobacco tins full of pins and bolts, rude postcards wrapped in bundles, ancient watches frozen in time, and faded sepia photos of dogs and horses, smiling women in hairnets and stern-faced men in trilby hats smoking pipes. One man's historical driftwood washed up by the tide of life and deposited in this desk. Needless to say, it was a source of endless fascination for me. Dad bloody hated me going in there. 'You've been in my desk again! Have you broken anything? Keep out of there! It's none of your business.'

There was one object, though, that fascinated me above all the others. It was a pint-sized eighteenth-century china beer mug that sat wedged in a forgotten corner on top of a bookshelf in the hall. This mug depicted what looked at first sight to be a serene oriental landscape. There was a willow tree, the branches of which hung down and kissed the ripples of a small stream, a pagoda, and, in the background, snow-capped mountains. Pretty standard stuff really. But it held a secret, this mug. If you looked very carefully, squinted your eyes a bit, and moved it away from you, all of a sudden an image of a man fishing would emerge from the background. He was made up of the landscape all around him: part tree, part pagoda, part mountain, and part stream. This man was hidden in plain view by a landscape to which he was integral. And that is what it is to be by the side of a river. Of course I understand that now. And I understand what Ratty means, too. But for so many years I was just an industrious little mole stumbling blindly through my environment, seeing but not watching, hearing but not listening. Searching but not finding.

Of all the birds there is none that embodies more this feeling of emerging from the darkness and into the dazzling light of an exciting and undiscovered world than the kingfisher. Whenever I see that flash of burnished orange and electric blue spiriting itself across the top of the water like a silent low-flying Spitfire at zero-degrees altitude my heart jolts. To see a kingfisher in flight is both incredibly exciting and incredibly calming.

The ancients believed that kingfisher nests drifted on ocean currents, spreading calm for miles around. The word 'halcyon' is derived from this myth about the kingfisher. The ancient Greeks knew it as the halcyon bird, which was said to have the power to calm the wind and the sea. Certainly on the days when I see a kingfisher I don't need family, friends or even a therapist to calm my troubled soul. On those days I need nobody because I have my halcyon bird and it bathes me in happy tranquillity.

Although the reality of the kingfisher's nest is somewhat less glamorous. Far from spreading calm while drifting serenely on ocean currents, their nests are actually squalid little chambers full of half-digested fish bones and other refuse located at the end of deep tunnels that slope upwards from the face of the riverbank. It's astonishing that such a wondrous miracle of nature can emerge from such lugubrious surroundings – like a supermodel living in a squat.

I don't always see a kingfisher when I want to see a kingfisher, but that makes the odd sighting here and there taste that much sweeter. The chief source of excitement comes in large part from this little sparrow-sized bird's extraordinary colours. Set against the subtle greens and browns and the grey skies of my native land, it looks like an exquisitely adorned visitor from the Amazon rainforest. It lights up the riverbank with its exoticism, like a dazzling samba dancer skipping through a dimly lit London pub. As rare as it is astonishing. There is also something peculiarly satisfying about the soundlessness of its flight. If I am lucky enough to see a kingfisher fishing, I'll watch, agog, as it spears into the water without making a noise or even a splash. It slices in like a razorblade. In fact, such is the aerodynamic excellence of the kingfisher's beak, it is said that the Japanese fashioned the noses of their bullet trains on it.

At the same time, or at least shortly afterwards, a great wave of peace descends upon me like a warm embrace.

And this became a very important part of my healing process after Mum died. I had already reaped the benefits of looking up and seeing house martins cartwheeling blissfully through the air or listening out for the soothing songs of the warblers and the thrushes – taking great solace from this abundance of life – but the kingfisher and its riverbank was a whole new world for me to discover. Here was a bird completely in tune with its watery ecosystem, an ecosystem that didn't so much provide individual moments of solace but that joined the dots of nature, that showed me how the natural world comes together as a single organism: an interlinked chain of life – of which I was very much a part.

Once I'd seen my first kingfisher on our local canal, I became an addict. I walked mile upon mile, hunting kingfishers down – trying to work out each individual's territory. One day, on a five-mile walk, I saw four kingfishers. It was a triumph. Quite apart from all those delicious jolts to my heart as I saw each one, set against the quiet serenity of it all, the knowledge that we had such a healthy local population of a bird that is in decline over so much of Britain was an added balm to the soul. These are my kingfishers and I know where to come to find them.

More than anything else, looking for those kingfishers taught me that I am never alone in nature. In particular, I am never alone on the riverbank. Although kingfishers are deeply solitary birds – nature's loners – and cannot stand the company of other kingfishers, which makes the mating season an awkward affair, their presence is integral to any healthy and diverse riverbank. Most especially chalk streams.

About five months after Mum's death, in late June, I went fishing with three old friends on one of the local chalk streams near my home in Wiltshire. I had been steadily building up my love of the birds around me, garnering inner strength from the daily blessings these birds bestowed upon me, learning lessons from them, but in a funny sort of way I saw each bird in isolation. I had yet to see the whole picture: nature living and breathing as a single organism. I had yet to truly feel the power of all the different elements of the natural world coming together as one glorious whole. Ratty's riverbank writ large.

Chalk streams are unique and precious ecosystems that harbour an abundance of life. At the source of these 'silent silver streams' the water literally rises up out of the ground – like a miracle – bubbling up from deep aquifers before setting its course for the sea sedately and without fanfare, over gravel and flint beds that purify and cleanse and make the water gin-clear. And because of the way these streams are managed, surrounded by lightly grazed water meadows – with no intensive farming – the landscape looks as it would have two or three hundred years ago, or earlier even. Standing beside the side of a well-managed chalk stream is like standing in a painting by Constable. It inspires and it heals in equal measure. And it was on this sunlit and scattered showery day in one of these heavenly places, and among these three close friends, that I had never felt quite so alone.

Despite the fact that I was moving on with my life,

and that with the help of my newly found connection to wildlife was finding much-needed perspective and reassurance, a profound feeling of loneliness gnawed away at my soul. I couldn't shake it off. The flood of broad-shouldered friends and family, heartfelt letters and sympathetic calls that had flowed around me just after Mum's death had receded to a trickle. Many months had passed and people had, quite naturally, moved on with their lives, but the fact of the matter was that Mum was still gone, Dad remained broken, I was still raw, and there was no one I felt I could talk to. And no one I really wanted to talk to either. I found it just so bloody awkward talking about it with friends.

Not so much for me but for them; I hated inflicting awkwardness on others. People just didn't know what to say after the slow-motion six-month car crash that had been Mum's diagnosis, prognosis and death. And if they did pluck up the courage to say something, I could sense their unease in case they might say the wrong thing. Then, of course, there was the intense awkwardness of encountering the people who didn't know. I felt particularly sorry for those people.

'Hello, Charlie, long time, no see! How's your lovely mum?'

'Erm, well, I'm so sorry to tell you but she died in January.'

Silence. Feet shuffling. Intense discomfort.

'Oh Lord, I'm so sorry. I'm so terribly sorry. I had no idea.'

I felt great pity for those people because I knew

exactly how they felt: the overwhelming shame that you *didn't know*, the guilt-riddled, totally irrational sin of ignorance. It was much easier, I found, just not to talk about it at all. So I followed the advice of one of my mother's oft repeated sayings: *Least said, soonest mended, darling.*

Pull yourself together, Charlie! For God's sake, get a grip, I told myself, as the river rippled and bubbled past me and a grey wagtail bobbed gently on the far bank. *Just look around you!*

My escapes to the river, with the same three friends, Jamie, Mark and Ninian, had become a bit of a tradition over the preceding years. They are three names that when put together in a sentence and preceded with the words 'popping for a pint with' elicit in Mary a minor convulsion. 'For God's sake don't drink too much,' she will say. 'And don't wake me up when you do get in!' But Mary also knows that these three friends are for me a pint-swilling pub therapy group, and so she tolerates my occasional misadventures with them.

They're the kind of friends in whose company you can completely relax. The kind of friends who will take the micky out of you mercilessly and tell you exactly what they think of you whether you like it or not. The kind of friends who'll pick you up when you fall — metaphorically and physically — and who'll drive thirty miles north of Birmingham on a rain-soaked midwinter night to pick you up when your car breaks down.

Outside the pubs and racetracks I regularly visit with these three characters, above all else I cherish those five or six hours spent pottering along the riverbank, fishing, chatting and drinking beer in cheerful riverside pubs. And in this particular June, of this particular year, Jamie, who organizes these jaunts, had chosen a conspicuously bucolic stretch of the River Kennet in Wiltshire not far from a decent pub. But no matter how hard I tried, no matter how beautiful the landscape or how sweet the birdsong, I couldn't locate my spark. My brain was like a concrete dam and it just wouldn't absorb any of it.

Seeing the river for the first time that day, calmly bubbling its way past me, and the water meadows with their drowsily grazing cows, willow and alder trees embedded in the landscape, that grey wagtail bobbing silently on the bank – the sheer damn serenity of it all – should have released in me a feeling of profound liberation. But not on that morning. On that morning I fought it and I remained, intently, a mole. It made so little sense. But nothing made much sense.

'You all right, Corbo? You're a bit quiet today,' said Mark.

Inside my head: *No, I'm not fine at all. I feel unutterably alone and I don't know why. I'm in one of the most beautiful places in all the world, surrounded by three of my favourite people, and yet I feel like I don't belong. And I cannot explain this feeling of dislocation from the human race in a way that you could possibly understand. Mainly because I don't understand it myself. I feel completely out of reach and stuck inside my lonely brain.*

Out loud: 'All good, mate. In fantastic form.'

I took the decision to focus on the fishing. To get on with the matter in hand. Wait patiently. And hope that somehow nature would work its magic, as it had done on so many occasions before.

Fishing is all about patience. And a good fisherman will always spend the first twenty minutes or so of the day watching the river with the keenest of eyes to ascertain what kind of bugs and flies are buzzing above the water waiting to be a trout's breakfast, so that he or she can mimic those flies on the end of the line. But I'm not a very good fisherman, so I just ask Jamie and he tells me what to do.

The thing I always find about fishing for trout on a fly is that there is the way I imagine the day might go and then there is the reality. In my head when I woke that morning was the beautiful chalk stream backdrop (so far, so good) with me casting a line elegantly and with precision towards a brown trout I had expertly identified under the water. The fly lands, oh so gently, atop the water – like a fly should – with just the right amount of delicate pflumph. The trout, with his dappled uniform of red, gold and brown, sees my fly – likes the look of it – and rises from the depths towards it, mouth gaping as his nose presses through the meniscus of that pure, crystalline water. My body tenses with nervous anticipation. He takes the fly. It's worked! *For God's sake don't mess this up, Charlie.* Strike! I have him on the hook. Intense relief. But now for the hard bit.

The trout tears off down the river with my line whistling off the reel. My heart flutters like there's a hummingbird in my chest, and the trout and I do battle. My line tightens and the rod arches gracefully, exactly as it should when you have a fish on the line. I reflect to myself how utterly absorbing fishing is, and how all my worries float away downstream as my mind focuses on the supreme challenge ahead of me right now of landing this fish. After five minutes, maybe more, the trout's fight is gone and he succumbs meekly to my efforts, gliding silently, serenely, into my awaiting net.

And then there is the reality. I get my line caught in a tree on my first cast. I lose the fly in the branches because I've failed to tie it on properly and startle every fish within 200 yards as I stagger about the riverbank, cursing, while I try vainly to untangle my line. After about twenty minutes of supreme concentration and effort, I have untangled the line and attached another fly, one which I hope resembles the flies darting above the surface of the river, and that some unlucky trout will want to make it his breakfast.

I'd like to think I am an expert at spotting the shadowy forms of fish moving under the water, perfectly in sync with their natural environment, and then having the ability to cast a fly directly above their noses, but I'm not; I'm hopeless at it. I'll wander down the riverbank with a keen-eyed fisher friend as he points out trout after hidden trout. 'Look, Charlie. Do you see that piece of weed on the far side of the bank, below the oak? There's a big trout the other side of it, tucked under the bank there. If you can just cast upstream of it . . .'

'Hmm, ahh . . .' Pause for effect. 'Ah, yes. Now I see. What a beauty!'

Nine times out of ten I've not seen a fish at all, but I don't want to let people down, so I lie, and then I thrash away vaguely at the bit of water my friend has pointed out, hoping that by some gargantuan fluke this particular trout, possibly senile and very hungry, will take my fly.

In line with past form I comprehensively failed to catch a trout that morning. None of us did. It was approaching midday and a steady drizzle had set in. Hardier fishermen than us would have continued, but we were anything but hardy fishermen. And so, in time-honoured fashion when four men have a decision to make, we decided to go to the pub for lunch. The rain, very conveniently, gave itself an air of permanence, which provided us with the perfect excuse to order more wine, and then some more. And then, just for the hell of it, one more bottle. Our little table in that cheerful riverside pub, with its belly laughs, booze and friendship is what the word 'conviviality' was invented to describe. And yet despite the cheerful atmosphere and good company, I was brooding and uneasy inside. I felt alone.

By the time we got back to the riverbank the sun had come out and fishing was the last thing on my mind. All I wanted to do was take myself off to a quiet meandering bank somewhere and be on my own. To sleep. To escape. I said to the others I was going to try a few casts upstream and headed off to a hidden place I'd spotted in the morning, where the river took a giant U-turn back on to itself and there was a cracked and broken old willow that could give me shelter should the drizzle decide to return. Using my fishing bag as a pillow, I lay my head down and very quickly fell into a wine-induced riverside trance.

Lying on that riverbank in Wiltshire I experienced for a few moments what it is like to be a living and breathing part of a wider whole – as integral to the scene around me as the willow tree swaying above or the grey wagtail bobbing peacefully downstream of me. Just for those moments, I was the man hidden in the serene scene on the old painted china beer mug. I was embedded in my own picture.

The bold yellow irises of the spring had evolved into the delicate pastel palette of pink, purple and blue of rapturously named wild flowers like willowherb, loose-strife and forget-me-not – all set against thick deep green and impenetrable clutches of reed. And what a racket that reed was making.

I could hear the powerful chattering refrain of what sounded to me like a cathedral choir of singing reed war-blers, although at the time I didn't know they were reed warblers. All of a sudden a moorhen scuttled out from under the bank beneath me – running across the surface of the water with a frantic dash, like its life depended on it, kek-kek-kekking as it went. Everywhere I looked, nature was imposing itself on my senses. Dragonflies droned lazily across the water. A grey heron, the great trout-guzzling adversary of the fisherman, assured in its own magnificence, evaluated me with its beady yellow eye before launching its long-limbed frame awkwardly into the air – a J-curve of flight – its giant wings flapping so slowly as to defy all the known physics of aviation.

What I had seen before as a lonely spot under this willow – a place to wallow in my solitary self – was any-thing but. And yet, still, there was one piece missing from this busy riverbank scene, one more colourful piece of nature's jigsaw I longed for to complete the picture. The kingfisher.

But I knew then it would come if only I waited long enough, had patience. And, as if on cue, there it was – the briefest of brief flashes of orange and electric blue darting elegantly across my field of vision. A split second and it was gone, no doubt scything into the water out of view after a tasty minnow or stickleback. But that was enough for me. The picture was complete. I rushed back downstream to my friends, bursting at the seams with news of the kingfisher. And while lunchtime's beer and red wine had deposited an inky black residue behind my eyes, I felt spiritually refreshed. Like the water around me, my mind was somehow cleansed and purified. I could see so clearly then, why the ancient Greeks called it the halcyon bird and believed it had the power to calm storms.

I came to accept that day grief is a lonely place and there is nothing I can do about that. Life is a lonely business, too. But I realize now that loneliness is part of the human condition and every single person on the planet experiences it, bar none. It is part of being human. While I remain no stranger to loneliness in all its forms, physical and metaphysical, I knew from that day on that nature could provide a way out of loneliness. I am Mole rolling in the meadow, released from his hole, alive to the wonder of his surroundings, and accepting that no matter how industriously I search for answers, it's unlikely I'll ever find them.

Kingfisher

What it looks like: *A sparrow-sized bird with an iridescent blue back and wings, burnished coppery-orange breast and a beak like a black dagger. It is an unmistakable sight as it darts across the river at low level. It is almost impossible to tell the male and female kingfisher apart.*

What it sounds like: *A high-pitched small shriek as it flies.*

Where to find it: *On the banks of rivers and canals all across lowland Britain. It nests at the end of long thin tunnels that it excavates out of the bank, and at the end of which it creates a small chamber for its eggs.*

What it eats: *Small fish, like minnows and sticklebacks, make up the vast majority of the kingfisher's diet. But it will also eat river-dwelling insects, tadpoles and shrimps.*

Chance of seeing one: *You have about a 25% chance of seeing one, with patience. Despite its bright plumage the kingfisher can be very hard to spot. The best way to come across one is to sit on the bank of the river or canal and just wait. Kingfishers like to find overhanging brances on which to perch and fish. Keep an eye out for the briefest of flashes of electric blue and orange as they speed along the surface of the water.*

11. Curlew

I take my gladness in the . . . sound of the
Curlew instead of the laughter of men.
                    Anonymous, 'The Seafarer',
                              approx. AD 1000

What, have wings and stay here?
                    Samuel Johnson

*On a still day one can almost feel the air vibrating with the blessed
sound.* I stopped in my tracks. I was reading my constant
companion, *The Charm of Birds*, sitting up in bed one
night, unable to sleep, and these words just leapt off the
page at me. It was Edward Grey describing the call of
the curlew on a spring morning. I laid the book down,
closed my eyes and imagined myself as a child on one of
our annual family holidays to the Isle of Mull, alone on
the hill, listening in wonder to that haunting, liquid, bubbling, otherworldly call flowing out across the grassy
tussocks. He goes on to write:

> The notes do not sound passionate: they suggest peace,
> rest, healing joy, an assurance of happiness past, present
> and to come. To listen to curlews on a bright, clear April
> day, with the fullness of spring still in anticipation, is one
> of the best experiences that a lover of birds can have.

Well, that was it. The answer to all my problems. I had to go to the Isle of Mull as soon as possible and track down a curlew. I didn't care so much that it was June and moving towards July (the end of their breeding season) and that it was pretty unlikely I'd hear the call of the curlew. But there was still a chance I would see and, just maybe, hear one.

The curlew, once deliciously known in some districts as the great whaup, is a mystical bird of the hills and coastlines of Britain. It is a wading bird that breeds in the uplands of England, Scotland and Wales in the spring and summer, and migrates south in wintertime to seek shelter from the cold and take advantage of the abundance of food on offer in the mudflats of river estuaries. Or, as the old folks used to say, the curlew is as well known to the shepherd as it is to the fisherman. It is a gracious animal about the size of a small rugby ball with elegant long legs and a mottled brown and buff uniform. It has a slender, long neck that gives it a dignified bearing, almost supercilious, and its downward curving beak is shaped like a fine sickle – its tip reaching gracefully towards the ground. All the better for finding a buried shrimp or two. And once it does find a shrimp or worm it flips it up into the air and catches it in its mouth.

Reading about the curlew resurrected in me a long-held desire I had had throughout Mum's illness and subsequent death just to bugger off somewhere on my own – to escape all of it – and to think hard, and without distraction, about what it was I actually wanted from my life now. I wanted the perspective that only boundless empty hills, and the haunting call of a curlew, could provide. Or, as I had it in my mind then, I wanted to 'do a Thoreau'. David Thoreau wrote a book in 1854 called *Walden; or, a Life in the Woods*. It was about the two years he spent in isolation in a cabin at a place called Walden Pond, which was in the (then) wilderness of Massachusetts in the north-eastern United States of America. He said of this experience:

> I went to the woods because I wished to live deliberately, to front only the essential facts of life, and see if I could not learn what it had to teach, and not, when I came to die, to discover that I had not lived . . . I wanted to live deep and suck out all the marrow of life.

I decided I wanted a little slice of that living deep and sucking out the marrow of life. To get away from the frantic hustle and bustle that was erupting again all around me. I wanted to feel for myself the healing sound of a curlew on a lonely moor.

Obviously a log cabin by a pond in a wild wood in 1850s America was not exactly realistic and, unlike Thoreau, I could barely spare a week, let alone two years. But I had enough time to get myself up to the Inner Hebrides and experience a bit of the uninhibited solitude that I so desired, and that I felt only the Isle of Mull could provide for me at that time. I wanted to re-establish a connection to the landscape and wildlife that had so enriched my time spent there as a boy. I wanted to feel again the sense of limitless space and possibility that that landscape had always engendered in me. And so I resolved that sleepless night to get in the car and drive to Mull at the first opportunity.

I had spent all my childhood summers on the Isle of Mull, staying with my grandparents on their farm. And, as a post-A level teenager, I had worked up there too. My grandparents died in the late nineties, after which my uncle took on the reins with his wife and family. The farm is situated on the rump of the island, far from the usual tourist trails, and is made up of a craggy, granite coastline, wind-battered slate-grey beaches and dramatic hills that sheer precipitously up from the sea. It is accessed by a single road that winds through the hills and then down alongside two sheltered lochs, one seawater and one freshwater, through a tiny cluster of stone houses – the village of Lochbuie – before coming to an abrupt end at the Atlantic Ocean. Or, as my Uncle Jim, Dad's brother, says with a glint in his eye when giving directions to visitors, 'Stop when you get to the sea.'

When it comes to farming, though, this landscape really is just a pretty face. Romantic notions cast aside, it is impossible to grow anything profitable on its thin soil, and so the only living that can be made as a farmer up there is through sheep and cattle, but mainly sheep. It is truly remote in a way that can be spine-chillingly haunting when the mist encloses you, or deeply uncomfortable when the Mull midges form a thick shroud round your head. And when the weather really sets in, in the form of frequent gales and rainstorms, the veneer of civilization is very quickly swept away. It is one of the last true wildernesses left in Britain.

When my friends were backpacking around Australia, expanding their minds on beaches in Goa or Thailand, or bungee jumping over Victoria Falls, I chose to be a shepherd's assistant on Mull and I lived in a glorified barn called a bothy. Not so glamorous as my friends in Goa and Bangkok, I'll admit, but just as rich in life experience. I had never been closer to nature. And it was this time that I remembered so vividly when I had originally decided to reconnect with the birds and wildlife. I'd reflected back then that in the months I had worked on Mull, I'd developed an intimate knowledge of the curlews and oystercatchers, buzzards and eagles, peewits and redshanks that lived all around me; birds that like the wild places and empty coastlines of Britain. I became quite close to what I had always wanted to be: *a good countryman*. Or, as one of the memorials to a long dead laird on the island reads, 'a useful country gentleman'.

I could gather a recalcitrant group of highland cows lost in the mist on a low moor and then walk them four miles home in a gale, treat a maggoty tup (don't ask), clip a wild blackface sheep with the best of them and reverse a tractor and trailer in my sleep. I was even taught how to slaughter a sheep for the table (not something I bring up too often with my vegan friends).

One memory stands out and it goes a long way in summing the place up. We were clipping sheep in a remote stone enclosure – what the Highlanders call a 'fank' – many miles from civilization and, I suspect, as far as it is possible to get in the UK from an electric power socket. But we'd managed to get electricity to this remote place via a diesel generator dragged five miles down a bumpy track on the back of a Land Rover. It meant we could use electric clippers for the first time, rather than hand shears – a hugely labour-saving innovation. My Uncle Jim – mid-electric clip – turned to one of the men working in the sheep fank that day and said, 'You see, James, Lochbuie is finally moving into the twentieth century!'

The Right Honourable Sir James Grant of Grant, the sixth Baron Strathspey, Baronet of Nova Scotia, thirty-third Chief of Clan Grant, looked up from the wool bag he was packing, slowly removed a Rothmans cigarette from his mouth, spat out a bit of wool, and said with a wry, nicotine-stained grin, 'Aye, Jim, just as the rest o' the world's moving into the twenty-first!'

While it was incredibly hard work, punishing at times, those five or six months after my A levels working on the Isle of Mull were as close as I've come in my life to feeling at one with my environment. I can still conjure an image of myself back then, aged eighteen, cocksure, driving a little too fast in a battered Series III Land Rover (that I would hit with a four-pound hammer to get started) down a remote and rocky track, no doubt a sheep bouncing around in the back. Atlantic rollers are crashing on to the beach to my left and a rocky hillside climbs steeply up towards the sky to my right. I was on top of the world. Master of my domain.

But the intervening decades, while fulfilling and enriching in countless other ways, had robbed me of that knowledge. It had robbed me of my closeness to animals, and most importantly to the solace that being near animals, not just the birds, brings with it. I wanted to go back and taste all that again, however briefly.

The family farm in Scotland was a menagerie, bursting at the seams with animals of all kinds. My family have always understood innately the benefits of being near to animals, both wild and domestic. They simply could not conceive of a world without them. And when I look back at all those different characters: assorted grandparents, aunts, uncles and cousins – some more eccentric than others – I can never remember any of them being excessively sad or depressed. Angry, stressed-out and shouty quite a lot of the time, but never fundamentally depressed. They were never haunted by any gloomy introspection or lingering reflections. They were experts in just getting on with things. Life to them was – and remains – something just to be lived. After all, there were chickens to feed, sheep to clip, fences to fix, missing dogs to track down. I have always envied that 'just-get-on' gene.

At the top of the tree was my grandfather, the patriarch: tall, sharp of eye and irascible. He had a face like a dignified, pipe-smoking buzzard – and was armed with a wicked sense of humour. He exercised this kind of magnetic power over all of us – we were at once in awe of him and at the same time in fearful trepidation.

Though this rather austere Edwardian figure could also be a rather naughty old buzzard. He was undoubtedly a charming fellow – a character – and well loved by those around him. I can remember him trying to bribe my sister ten pounds to get a tattoo and my brother twenty pounds to get an earring – just to irritate Dad.

He'd take great pleasure in teaching us all very off-colour jokes, and loved it when I drew rude caricatures of people (including him). And he adored my mother. We all knew it was because she was the only person in his entourage that stood up to him. She took no flack from the old buzzard at all.

You could not find a more beautiful location for my grandfather's house in all Scotland. It was serenely situated, facing south-west, just yards from the sea in this remote corner of paradise. It had an unimpeded view across a spreading bay, embraced on either side by glimmering hills that seemed to stretch out towards America and that would turn golden yellow with gorse blossom in springtime. Its name, Lochbuie, is Gaelic for yellow loch. No wonder I wanted to go back.

But inside the house it was a different story. I'm not sure my grandfather had spent any money on it since, well, ever. 'I wish this place would fall down and I could live in a nice warm bungalow,' he said to me once. And I don't blame him. There were buckets in the bedrooms to catch the water that seeped through holes in the ceiling when it rained, and here and there would be little signs propped on chairs that read: *Do not walk here. You will fall through the floor.* It was profoundly damp. And even in August I would sleep at night in my pyjamas, dressing gown, woolly hat and socks.

And yet despite all the discomfort and damp, as children we didn't really mind. This big crumbling house – with its austere beauty – was a source of daily adventure and wonder for us: cavernous empty rooms full of furniture you could climb into, long dark corridors that led to intriguingly locked doors, and it was imbued with any number of ghost stories that would make my toes tremble. I was never bored. Especially with all those animals around, many of which were stuffed. Ancient corpses of long-dead seals, owls and eagles (remnants from a wilder, less sentimental age) would peer down at me with their glazed eyes from great glass cabinets dotted around the house. And all those antlers! I once counted forty-two sets of red deer antlers on the walls of the dining room alone.

And there were the live animals too, of course. Chief among these creatures in my memories of the place was McTavish, the tame blackface ram. Well, I say 'tame' in the loosest possible sense. McTavish lived in the hall of the house and would lurk menacingly near the front door all day, waiting to pounce. To my young and fearful mind he was a malevolent sheep, short-tempered and with giant horns, who made getting into the house a high-risk activity. If you were lucky, he'd be looking in the other direction and you could sneak by, but if not, you risked being tossed over his head like an inconvenient rag doll. My aunt used to take him shopping with her.

My grandmother, resilient and kind-hearted – all tweed skirts, hairnets and 'yoo-hoos!' – at one point adopted an orphaned red deer, which she called Babycham. It used to live in the house (like McTavish) and lie at her feet in the sitting room while she read *The Racing Post* by the fire, snuggled up next to her two West Highland Terriers, Midge and Meg.

But uppermost in my memories of that time are my

grandfather's two Yorkshire terriers, Shuna and Mini. They were a particular menace right up there with McTavish the cantankerous ram. Perhaps the biggest potential crime any of us could commit while under my grandparents' roof – something that kept my father awake night after cold, damp night – was to sit on, tread on, run over or by some other means – accident or design – kill one of Grandpa's dogs. They were excessively hairy, violent little brown and black creatures that bit ankles hard and bit them long. They were like an angry pair of bedroom slippers. We lived in constant fear of them. Not just because of their ankle-shredding tendencies, but because their size meant they were just so easy to tread on. Richard's Jack Russell terrier, Fly, was banned from my grandparents' house for the very reason that she would immediately mistake the Yorkshire terriers for rats and that would be the end of them and, more worryingly, us.

Needless to say, my grandfather adored these menacing pint-sized hounds. They wielded immense power in the household. They slept in the bed at night and would usually spend most of the day peeking out from one of Grandpa's jacket pockets, growling at anyone with the temerity to get too close. I can still picture my grandfather's towering, elegant frame, his ancient tweed jacket hanging loosely off his broad shoulders, with a dog in each pocket. Double-barrelled.

We never did do any harm to these dogs, by accident or design, thank goodness. But it did mean you could never sit down with full confidence. When not in a bed or a pocket, they would disappear chameleon-like into the nearest armchair. The best piece of advice on life my father has ever given me, was this: if ever you *should* accidentally kill one of Grandpa's dogs, for God's sake don't tell anyone. Just bury the body and join the search party.

In addition to the dogs, deer and sheep, there were two doves called Emily and Phoebe that at one time lived free range in the house and roosted in a lampstand in the sitting room. There were chickens that laid eggs on the sofa, alongside the peahens, and in my grandfather's younger days, so I have been told, he kept a tame fox that went everywhere with him, a monkey called Bimbo and a small pony that would often travel with him in his car.

All these memories and family stories crowded into my mind as I drove up the backbone of Britain on that late-June day on my way to find a curlew. I never cease to be amazed by their eccentricity, these blood relations of mine. I could see then so clearly why my mother who, in comparison with these monkey-owning, fox-taming loonies, came from a deeply orthodox background, fell in love with it all. And all that eccentricity ultimately produced my father. I loved so much as a child being part of it all. I was just so proud. And I knew, even then, how lucky I was that my summer playground was this magical, endlessly intriguing and remote Hebridean island.

It takes twelve hours to get to the family farm on Mull from where I live: 500 miles across England and Scotland, an hour on a ferry and then another fifteen miles on the island. It's easier, in fact, to get to Hong Kong.

Once the motorway runs out at Stirling Castle, the road snakes very quickly up into the Highlands, through deep-cut glens and alongside shimmering lochs – and you can sense the maddening crowd disappearing further and further behind as you go. It runs through places with whimsical names like Crianlarich, Dalmally and the Bridge of Awe, and then all the way down again to the scenic little fishing port of Oban in Argyll. I must have done that drive a hundred times and I have never yet tired of it. More than anything else, it gives me plenty of time to think. And then at the end of this journey of long and winding roads is a sea crossing.

I still feel the same shiver of excitement when I see the black, red and white livery of a 'Cal-Mac' ferry steaming into Oban Harbour. Even as I write this, in a small bedroom in rainy Wiltshire, I can taste the crisp, salt-inlaid air, infused with the pungent scent of that day's catch from the Oban trawlermen. There is something about getting on that ferry that I find incredibly healing. I leave my problems on the pier at Oban and watch them disappear into the distance as the boat carries me to my place of peace. It is a hauntingly, staggeringly, mind-alteringly beautiful corner of Planet Earth. And I think some of its peat bog-infused heathery aroma must have seeped into my genetic code.

I arrived at my destination, a small farmhouse that belonged to my father, located just a few miles up the lane from my grandfather's old house, at about seven o'clock in the evening. I parked the car, dumped my bags in the bedroom and set off immediately for a walk down to the loch. The sun was setting with a benign ease behind the hills to the west, turning the sky a comfortable shade of pink and casting friendly shadows across the bay. I was keen to make the most of it. The Isle of Mull has one of the highest levels of rainfall in the whole of Europe, so if the sun does decide to shine, you don't waste it. My chief excitement, though, as I made my way down to the loch, was the prospect of hearing that curlew; it would have completed the picture. It didn't call. Not that evening. No matter. I still had plenty of time.

The next three days and nights were punctuated by, well, nothing at all. Gloriously, nothing at all. I settled into a routine of long walks during the day, fireside whiskies in the evening and heinously early nights. In fact, I had not slept so consistently well in almost a year. Deep, dreamless sleep of the kind I had longed for. But no curlews.

One of the things that I have always loved most about the west coast of Scotland is the overwhelming feeling of decay. Not just in the very physical sense of rotting vegetation and salt-air-corroded and weather-beaten buildings and machinery. That too, of course. But also the overwhelming sense of lives lived before, long gone now, but echoing back to me through the past.

As a child on Mull I had always felt surrounded by this feeling of eerie and yet strangely life-affirming decay. Because there has been so little development on the island over the centuries, and so few people live there, you can see the residue of past generations standing out in the landscape all around you – going back tens, hundreds and a thousand years or more. You can read the island's history in its hillsides and shorelines. From the Neolithic stone circle just yards from my grandfather's house, the ruined fourteenth-century castle commanding the bay, shrouded in a cloak of ivy, to long-abandoned houses and steadings with the roofs caved in, trees growing out of the walls and sheep grazing where a kitchen might have once stood. Nature reclaiming its own. Cold, empty hearths sitting under crumbling chimneys that would once have glowed with family life. A disused pier – once alive with boats and livestock and the noisy energy of busy West Highland life, silent now, and collapsing gently – year by year – time-lapsing into the sea, the rotting skeletons of dead fishing boats scattered around it.

I loved, and still love, being surrounded by all this living and dead history. I was particularly drawn to the old machinery from bygone days. Rusting old tractors and trucks from the thirties and forties, agricultural implements half buried in the ground, abandoned. Someone had parked them in this spot all those years ago and just walked away. They'd lain unmoved ever since – and forever more – waiting to become fascinating ancient relics, long after I'm dead, and dug up by some future generation. The detritus of a busy agricultural life going back a thousand years.

One vigorous morning I decided to head up to an abandoned village I remembered high up on the hill and overlooking the sea. The west coast of Scotland is littered with these abandoned villages. Some of them were 'cleared' by unscrupulous landlords in the eighteenth century, the people replaced by sheep, while other villages died more lingering deaths, as the people gradually leaked away to the cities of the Scottish mainland or the New World in search of a better life.

My walk took me past a kind of panoply of decay: past the disused pier and roofless ferryman's cottage, up on to the hill, where I took my bearings from a ruined farmhouse, intriguingly known as Gortendoil – or blind man's field – and then headed south-east towards where I had remembered the village to be. It is, in actual fact, two villages next to one another and made up of a couple of dozen houses. There would once have been a thriving community of several hundred people up on this hill. But no one has lived here in one, two, three living memories, and the houses lie empty and open to the elements – row upon row of jutting stone teeth.

As I wandered through the village, sitting atop this gentle hill and with the Inner Hebrides reaching out before me, I reflected that this place, which felt to me like it had been abandoned forever, would have been inhabited by people for so very much longer than it has been uninhabited. All those unnumbered lives, countless stories, the tragedy and joy of ordinary life, was embedded in the stones around me. Silent now.

Mum and I had regularly walked here in my childhood. She would understand exactly what I was feeling, up here and alone. I know people go on all the time about how short our tenure on this earth is in the great scheme of the universe, but standing in a place like that brings it home in a way that actually has meaning. And I don't mean in a frightening oh-bloody-hell-my-time-is-running-out kind of way, which precipitates red sports cars and excessive skydiving, but in a reassuring we've-got-all-the-time-in-the-world kind of way. When I stood among those abandoned homes, looking out at the very same view that countless generations would have looked out upon, I didn't feel sadness or loss, or an overwhelming urge to create a bucket list, I felt like all my worries, anxieties and strains no longer mattered very much, not really.

Cue the curlew.

Where are you? Now is the time for your haunting *cur-lee* bubbling out through the ruins. If I were in a film or a Thomas Hardy novel, you'd definitely be calling, because, after all, *A moor without a curlew is like a night without a moon.* But no call came. I had the meadow pipits instead. Which was just fine. And, below, by the sea, I could hear the jubilant notes of an oystercatcher's sharp song, no doubt superbly adorned in his black-and-white tunic, red gaiters and lavishly orange beak. If a moor without a curlew is like a night without a moon, then a beach without an oystercatcher is like a sky without the stars.

After a couple of days of all this solitude and no curlews, Richard and Katie turned up at the house a little unexpectedly with their families. I wasn't sure quite how to take this. It rather buggered up my solitary lonely Thoreau act. How could I go around sucking the marrow out of things with all this lot invading my peace? But there was nothing much I could do about it. And it turned out that I wasn't really cut out for the Thoreau thing anyway. I'd only gone three days alone already and was beginning to tire of my own company – and I still hadn't heard or seen a blasted curlew. From a singular one the household grew overnight to a multitude of nine. It was a joy to be surrounded by these happy and noisy families. My people. Mum's people. We had a couple of days together, walking and talking – drinking and staying up too late – and we even visited the empty village up on the hill together. My four nephews and nieces galloped through the ruins merrily, and my brother and I speculated earnestly on how these people must have lived all those years ago, while the rest of the party soaked up the view.

Then we talked of Mum. She had so adored this island. And had put her heart and her soul into resurrecting a rapidly decaying farmhouse that had at one time been lived in by a shepherd but long been unloved and destined no doubt for a similar fate to the houses all around us. She turned it into a place of comfort, warm energy and laughter – the great project of her later life. A place where her children could come together in this wilderness we all loved so very much. It was the very farmhouse I had been staying in for the last week, tucked into the hillside overlooking the loch: solidly built, white-washed and with a kind of chaste beauty to it.

Shortly after Mum died we buried a rock into the hillside next to the house, overlooking the loch, and inscribed it with her smile. Or at least as close as we could get to it with these words: *A smile costs nothing but gives much. It takes but a moment, but the memory of it sometimes lasts forever.* And so it is too with the call of the curlew. It was time for me to go home. No curlew sang for me on that trip, I am sad to say. But it turned out I hadn't needed one after all.

About a month after I got back from my failed impersonation of a romantic naturalist, around the end of July, Mary and I went to visit her granny in the little seaside village of West Wittering in Sussex, just over an hour's drive from our home. West Wittering is Mary's own place of peace, and where she spent all her childhood summers with her grandmother – over a hundred years old by now, with sixty-two living descendants, her family's own magnetic, all-seeing matriarch. To stay with 'Granny-by-the-sea' at her house nestled into the side of the Sussex coast was to find a haven from whatever real-life struggles might be overwhelming you.

It is a place that has special meaning for Mary and me. Because it was there, under the branches of a sprawling ancient oak down by the side of the estuary, drowsy with wine and surrounded by a flurry of May bluebells, I had proposed to Mary and she had accepted. Every time we visit West Wittering we make our way to this very same spot, overlooking Chichester Harbour and a short walk from a pretty little pub called the Ship Inn. On this day we were walking the very same well-trodden and happy path, talking of this and that and nothing at all, when I was stopped in my tracks.

'Mary, Mary, listen! Can you hear it?'

After all those miles and all that searching, what

should I hear echoing hauntingly, joyously, across those southern-English mudflats just an hour and a half from our back door? Why a curlew, of course.

> The tide rises, the tide falls,
> The twilight darkens, the curlew calls;
> Along the sea-sands damp and brown
> The traveller hastens toward the town,
> And the tide rises, the tide falls.
>
> Henry Wadsworth Longfellow

Curlew

What it looks like: *A large wader with mottled brown and white feathers, long legs, a slender neck and distinctive downward-curving beak.*

What it sounds like: *The call of the curlew once heard is never forgotten. It makes a distinctive, liquid, bubbling 'curl-leeee' sound that echoes out across lonely moors and upland areas in spring and summer.* It's as if solitude has found a voice, *as I once saw written.*

Where to find it: *Curlews breed in upland areas in the spring and summer, making nests on the ground in shallow depressions. They favour tussocky grassland and heath, full of invertebrates and where they can best camouflage their nests from egg- and chick-eating predators. Outside the breeding season, in winter, curlews retreat to the coast for warmer weather and to feed on the abundance of food in the mudflats of estuaries.*

What it eats: *Curlews feed on worms, insect larvae, spiders, caterpillars and other invertebrates when breeding in the hills. When they are on the coast in winter, they use their long bills to probe for small fish, crabs, shrimp, worms and snails.*

Chance of seeing one: *Of all the birds in this book, the curlew is the one nearest to extinction in Britain. It simply cannot cope* with the twin pressures of modern farming techniques and heavy predation from ever-larger populations of foxes and crows (which feed on the eggs in their ground-level nests). The UK population of curlews has fallen by over 60% since 1969. But despite this dramatic fall in numbers, it is still possible to see and hear a curlew with relative ease — for now. Winter is a good time to head to a coastal reserve to see one; this is because the native population is boosted by many thousands of migratory curlews, who fly down from northern Europe for warmer weather and more food. In the spring, heading into the hills to their breeding grounds — mainly in the north — is the best way to hear the curlew's unique and moving call — in particular on well-managed grouse moors.

12. Barn Owl

Joy cannot exist without sadness. Relief cannot exist without pain. Compassion cannot exist without cruelty. Courage cannot exist without fear. Hope cannot exist without despair. Wisdom cannot exist without suffering. Gratitude cannot exist without deprivation. Paradoxes abound in this life. Living is an exercise in navigating within them.

Julie Yip-Williams

The solemn temples, the great globe itself,
Yea, all which it inherit, shall dissolve,
And like this insubstantial pageant faded,
Leave not a rack behind. We are such stuff
As dreams are made on, and our little life
Is rounded with a sleep.

William Shakespeare, *The Tempest*

Alake, alake, when the clock strikes twal,
What soun's an' what sichts are there;
When the howlet flaps wi' an eerie cry,
Through the woods o' Knockenhair!

Alexander Anderson,
'The Deil's Stane'

In my end is my beginning

Years have passed since Mum's death. I am driving over the roof of Wiltshire, east to west, through the Vale of Pewsey, on a clear, undulating road. Around me is a glorious autumn sunset. Next to me is Mary. And behind me, our two boys: Arthur and Teddy. We're on our way back from my father's house. The car is full of laughter and the excessive noise that only two boys under the age of five are capable of producing, or at least these two anyway. The future is unwritten. I am happy. We are happy. And life tastes sweet.

'WATCH OUT!! For God's sake, Charlie. What the hell are you doing?? You could have killed us.'

'But it's the barn owl, darling. Didn't you see it?'

No, Mary had not. She was too busy gripping on to whatever she could, as the car careered across the (fortunately for us) empty road. The children had gone awfully quiet.

In my defence I'm usually a pretty good driver: diligent and careful. But not when it comes to spotting barn owls on autumnal evenings while driving down country lanes. The Highway Code and health and safety tend to go out of the window when I see our local barn owl. And, I'm embarrassed to say, I have form, Your Honour. See also: nearly going under a truck having spotted a murmuration of starlings on the M32 near Bristol back in November 2014 and veering wildly into the hard shoulder in February 2016 when surprised by a deceit of lapwings mobbing just after the turn-off for Newbury on the M4.

What I had seen to cause this minor motor wobble, which had left Mary ashen-faced and a year older, was the ghostly form of a barn owl swooping silently in the meadow to my right, hunting field mice and voles in the gloaming. The barn owl is perhaps one of nature's finest works of art. And I am lucky enough to have a pair of them that live not three miles from my home. They nest each year in a lonely Victorian barn that sits by the side of the elegant ribbon of Wiltshire road we were driving down that evening. What makes me even luckier is that once they have settled in a place that suits them, barn owls will remain all their lives. It is a happy thought that, barring the complete destruction of what is left of our local meadows, my family and I will have the pleasure of generations of barn owls to come.

It is a truly heart-stopping sight to watch a barn owl hunt at sunset on a warm September evening. What stands out, apart from the almost supernatural ghostliness of its pure white, brown and buff appearance, is the profound noiselessness of the whole affair. Barn owls make no noise at all. And I don't mean are mainly quiet with a hint of a 'swoosh' if you listen carefully, but utter silence. They are nature's silent assassins. Equipped with a face, feathers and a frame honed over millennia into the countryside's most devastatingly efficient night-time killer. Their heart-shaped faces direct high-frequency noises to their ears so that they can hunt by sound as well as sight, and their eyes are twice as sensitive to light as human eyes. In fact, it is thought that 75% of a barn owl's brain is devoted to its hearing and vision. You really, really don't want to be a small vole scrabbling about at night in the territory of a hungry barn owl. He'll hear your little heart beating. And rip it out. Watching a barn owl hunt makes you feel that you are in the presence of a creature that is truly at the very pinnacle of its game. One of nature's true professionals.

But what I love the most about barn owls is how beautifully economical they are with their energy. Some might even dare to call them a little bit lazy really. When not in their three-month breeding season, when a single owl can kill up to a thousand mice and voles, a barn owl will roost for up to twenty-two hours a day. They are nature's equivalent to the couch potato who doesn't leave the sofa all day except to pop out to the corner shop for a pint of milk and a can of beans before bed. And why wouldn't you sit on your bum all day if you were that damn good at your job?

And yet, like everything in life, nothing and nobody is perfect. Even the barn owl, with all that elegant perfection and lethal efficiency, is flawed. In its rush to design the quietest killer of all the birds, nature forgot to make barn owls waterproof. They cannot hunt in the rain. And if it is raining all through the breeding season, it can prove disastrous. But it wasn't raining as we drove back from my father's house on that sedate magenta evening, and the barn owl's soundless flight, pin-drop-in-a-busy-pub hearing and razor eyesight was on display for all to see and wonder at.

Despite the intense calm that watching a barn owl hunt inspires, and the feeling of inner peace it can instil, owls throughout history, and in all cultures, have been seen as bad omens – or even a fearful portent of death. Shakespeare's plays are littered with doom-laden owl references, like the murder of Duncan in *Macbeth*:

> It was the owl that shrieked, the fatal bellman,
> Which gives the sternest goodnight.

In the Middle East the owl was once regarded as a god of death, the embodiment of evil, and would even carry off children at night. Much of this fearful superstition, I think, can be laid at the door of the barn owl. Not just because of its ghostly presence and stealthy movement but because of the dreadful noise it makes when it calls. Not the merry hooting and nightly reassuring tu-whit and tu-whoos of his tawny owl cousin. Oh no. The barn owl's call is a piercing shriek. It sounds like someone is murdering a small child, so it is no wonder that the old name for it was the screech owl. No wonder either that people thought hearing it portended death or disaster. Right up to the 1950s people would nail barn owls to their doors to deter evil spirits, and in some parts of Britain the shriek of a barn owl was believed to foretell a big storm or the onset of cold weather. But on the positive side of the ledger it was also said that if you heard a barn owl call during a storm, then it was a sure sign the storm would shortly abate.

But shrieking, death and bad omens aside, the characteristic that owls have been most known for through history is wisdom. It has always been said of owls that they are wise. In fact, the owl's association with wisdom first arose back in ancient Greece. The goddess of wisdom and reason, Athena, was depicted with an owl on her arm, as was the Roman equivalent, Minerva, and this association has lasted millennia. But the truth is that owls are no brighter than any other bird. The myth persists, I think, because of their big all-seeing eyes, acute hearing and the ability to turn their necks almost 360 degrees.

When scientists have conducted intelligence tests on birds, it is the corvids – ravens, jackdaws and crows – that regularly come out on top in terms of reasoning and problem-solving, and owls are somewhere near the bottom. But what is wisdom anyway? The owl is perfectly suited to its environment; it doesn't need to go around solving problems. Most of its brain is dedicated to hunting and so it hunts. Very successfully. And then it roosts for twenty-two hours. Is that not wise? I wish I spent more time just getting on with the matter in hand rather than desperately trying to solve problems in my head.

Is wisdom, in fact, just having the knowledge that you don't have the knowledge? And being happy with that. Accepting the fact that there are no answers. Life is, perhaps, like quantum mechanics. 'If you think you understand quantum mechanics, then you don't understand quantum mechanics.' Or as an owl might say: 'I will never be able to recognize shapes, count left to right, use a tool or open a Perspex box with a nut in it, like my clever corvid cousins, but I'm at ease with that, thank you. Instead I'll just do what I do best and go and slaughter a vole for supper.'

Is it wise to understand that death and change, anxiety and depression, are as inevitable as rain on the opening day of a Test match in May? Is it wise to understand that despite all this sadness, life will get better again? If we just wait. Is that wisdom? Is it wisdom to know that you won't find wisdom on an iPhone or on Instagram? Or that the answer – on occasion – might just lie in putting that phone down and taking a long walk in the woods. Is it wisdom not having constantly to ask what is wisdom?

Some days I see my existence with a new set of large clear eyes – 360-degree vision. It dawns on me all of a sudden quite how fulfilling and dappled with joy my life actually is. Like being at the top of the hill in Mull, enclosed by cloud, when without warning the wind blows the mist away to reveal a shining vista of heather-capped hills, blue skies and an aquamarine loch. I remind myself that the vista is always there, and it is only the mist I create in my mind that obscures it. And it is in those moments, like that time in the car with my family, that I have an overwhelming desire to hop into a time machine and go back to all my various different selves and reassure that anxious and sad fellow that all will be well:

To the overweight and spotty undergraduate laughing and joking at the party but feeling so utterly unworthy, insecure and alone in a crowd, pining for acceptance:

'The zits will go, Charlie, don't worry. As will the weight. And all these people around you, the ones you think are so confident and assured of themselves, feel just the same as you. One day you will once again recreate that feeling of supreme confidence and oneness with your environment that you felt as you bumped along that rocky Mull track in the old Land Rover.'

To the underweight twenty-something, lovelorn without direction, a dearth of self-esteem and hopelessly out of his depth as he endures his first London winter:

'You're going to meet the love of your life one day soon, so don't sweat it. You'll find a place and a direction of travel. Just being you is good enough for now. Have faith.'

To the grieving thirty-something with all his lost joy, debilitating panic attacks, unfathomable melancholy and dread for the future:

'Life's a bitch, Charlie. And then you die! And one day that bumper sticker will inspire reassurance and a wry nod of the head and not crushing anxiety.'

I would even console the distraught fellow who, just months ago, had watched the magpie destroy the mistle thrush nest:

'They're back! The mistle thrushes are back, Charlie. I saw them only yesterday, chasing a rather harassed-looking woodpecker around the garden.'

For some reason, though, driving over that Wiltshire hill on the way back from Dad's house, and encased by all that loving security – and inner peace and barn owls – it was none of those past selves I wanted to speak to. I wanted to speak to a little boy – the eight-year-old me – standing alone by an orchard gate in Hampshire.

It is September and the apples are ripe, weighing heavy on the trees all around him. The holidays have drawn to a close and, after nine long weeks at home, he's a few days away from going back to school.

I want to put my arm round that little boy, like he was my own son, and reassure him that everything will be all right. That he has nothing to fear, not really. I feel so desperately sorry for that child because I know that he doesn't like himself very much. And he wonders how if even *he* doesn't like himself, why on earth would anyone else? I want to tell him that this feeling is utterly without foundation. That so many of us feel like that at one time or another, from his great heroes Stephen Fry and Hugh Laurie (this little boy adored *Blackadder*) to Agatha Christie, John Betjeman and Winston Churchill. For goodness' sake, Charlie, even the great Captain Haddock gets pretty low at times. And Tintin too, no doubt. (Though in the case of Haddock that might have something to do with the whisky.)

I would tell him that he is just a normal child, a flawed human like any other – stuffed to the gills with inexplicable sadness, inbuilt weaknesses and endless and unfathomable contradictions. I would say, 'That is the sweet mystery of life, Charlie.

'You're going to have so many happy times in the years ahead, too: real and tangible joy, and lots of it, to match all the inexplicable self-doubt and sadness. You can't have one without the other; they arise mutually, as the Buddhists say, like the bumblebee and the flower.' And I would wish so very hard that this little boy would believe me when I said it. I'm pretty sure he wouldn't, though. Cynical sod.

Some people are lucky enough to retain their faith in God. I am no longer one of those people or at least I don't think I have faith any longer. If I do, I am like John Betjeman, hanging on to my faith by my eyelids, while mainly thinking it's a load of old rot. And I think that most people no longer have an unquestioning belief in a fatherly spirit in the sky who will help us out of a fix and let us through his gates when we die.

Losing faith in a father figure in the sky was a big problem for me, not only because I had felt my religious faith quite keenly for a time at school but more importantly when it did start to dissipate I had nothing with which to replace it. At the time of Mum's death I needed answers and I needed them fast. I needed to find another faith. Science was no good to me. Science is all cold hard fact and challenging theses. Supremely important in humankind's quest to get to grips with Planet Earth and answer life-changing questions. But when I faced a personal crisis on the scale of Mum's death, or inexplicable anxiety and melancholy, science was never going to light my soul. It would never warm me on those days of merciless reality. Science is all about finding answers to questions. But what if my question doesn't really have an answer? Or what if I don't even know what the bloody question is?

It's so easy to be wise after the event. The real trick I've found is in finding a way *not* to require some irritatingly cheerful future version of myself to prattle on at me about how everything is going to be OK when all

I'm feeling is black despair. I've learned that life does get better again. If I just wait. And then it will come again, the darkness. And then it will go away again. And then it'll come back again. Good days and bad. Can't-get-out-of-bed days. Infuriating days. Angry days. Inexplicably cheerful days. Days when suddenly, and quite out of the blue, everything seems to make sense, like that day on the way back from my father's house.

Every day since I discovered that solitary lark on a Wiltshire hillside, I have been learning, more and more, about the power of the birds and the beasts and the landscape they inhabit to heal me. My barn owl, for example, taught me that nature is unsentimental. That you cannot have beauty without cruelty. Like the sparrowhawk snatching the starling out of the pulsating murmuration in front of my very eyes, or the explosion of feathers as a peregrine falcon smashes into a pretty little wood pigeon.

It can feel some days like the species to which I belong is suffering from some kind of mass disorientation, a collective anxiety attack. It can feel like we've forgotten how to live. So we look for answers on social media or in alcohol, red sports cars, bigger houses or in countless other dead ends. And yet, like the pot of mustard you've spent twenty minutes looking for in the cupboard, nature is in front of our very noses. And we've ignored it for too long. I had ignored it for too long. I had lost my sense of wonder at the natural world around me. Worse, I had lost my sense of reverence. And without reverence we have nothing.

I could no longer see, or wasn't particularly interested in, the awe-inspiring beauty of a glimmering yellow celandine on a springtime verge. A tiny goldcrest perching elegantly on the branch of an old oak. Or the blurry reflection of a Wiltshire hillside in a puddle. I'd lost my ability to see and to appreciate the sheer indefinable splendour of the everyday. And yet it was all free and on my doorstep. All my twelve birds can be found within an hour's drive of anywhere in Britain or Europe. In fact, most of them can be found within minutes from any door anywhere. We all just need to look up every now and again.

This book has not been an attempt to instil any great philosophy by which you can live a better life. I don't pretend I've got the answer. Or even the question for that matter. I am not an expert in birds either. Not really. And I'm certainly not an expert in mental health. But I am, perhaps, an expert in being me: a flawed human.

I understand more clearly than I have ever done before that life is a cycle, and my mother's death was an integral part of that. Nature took her away from me and it certainly won't bring her back, but via nature I also found a way through my grief. It was the birds and the landscapes they inhabit that showed me. I am *of* my environment and not merely in it. And I wanted to share that because I know it will make you a more resilient human being. And even, dare I say it, happier.

I can look back at the totality of my life and ponder endlessly about whether I was happy or not at different times. Sometimes there are clear markers, like meeting and marrying Mary, or the birth of my sons, but most of the time happiness is almost impossible to pin down. And the truth is, I have no idea what the secret to a happy and fulfilling life is. But I do know one thing for absolute certain: I know what it is to feel that there is nowhere in the world I would rather be than right here, right now. And that is when a robin perches on a fence post next to me as I burn leaves in my garden on a November afternoon, when I see a wren flit by on a crisp winter's day or I hear a song thrush sing out unexpectedly on the longest night. It's a busily intent bullfinch whistling softly in the lane or a clattering of jackdaws chackling on an autumn evening. It's seventeen sparrows bathing in a puddle outside my door. The first chiffchaff on a brutal March day, or the arrival of the house martins in May. The flash of a kingfisher sparkling on a June riverbank, a curlew's call echoing at the break of day across a Sussex mudflat or a barn owl hunting for field mice at the end of the day. But most of all – most of all – it's a skylark singing on an empty hillside.

This is the sound of my universe. And I am grateful for it every day.

Barn Owl

What it looks like: *A ghostly white presence when hunting silently at night. A purely nocturnal creature, the barn owl's upper parts are a mottled grey and hazel while its lower parts are bright white. It has a distinctive heart-shaped white face.*

What it sounds like: *The barn owl's call, unlike its more gentle tawny owl cousin, is a sharp and ear-piercing shriek uttered at the dead of night.*

Where to find it: *Barn owls nest on high flat ledges in disused farm buildings in and around grassland and meadows. They are distributed throughout Britain, but due to intensive farming they are often confined to hunting on pesticide-free field margins and the verges of roads, where they can still find their prey.*

What it eats: *Barn owls eat small mammals like mice and voles.*

Chance of seeing one: *While the barn owl still has a very wide distribution across Europe, it is yet another victim of modern demands for intensive food production. It is unlikely you will see one on a normal walk, firstly because they only hunt at night, and secondly because there are very few large wild-flower meadows — abundant with mice and voles — left in Britain. Or indeed disused barns for them to nest in. Many farmers now provide nesting boxes* *for owls, and it won't take much asking in a rural area to find out where a local barn owl is hunting. If you go out at night near a barn owl's nest, in the breeding season in particular, you will have an 80% chance of seeing or hearing one.*

Gazeteer

A year in the life of birds

When I first started to rekindle my knowledge and love of birds, the thing that frustrated me the most was that all the guidebooks tended to be stratified by species. Whenever I saw a bird that I wanted to identify, I would be forced to start at page one and work my way through the book till I thought I had found the right picture of it. But even then I could not really be sure, and I could not go back to the hedgerow or wood where I'd seen it to double-check. What I longed for was guidance on what I would be expected to see in certain places and at certain times. I needed context. For example: if you are sitting in your garden on a summer afternoon, it is very likely that you will see a robin, a great tit, a chaffinch and a greenfinch. If you look up, you also might see swallows and house martins swooping majestically about, catching flies in the air. If you listen, you might hear a blackbird or song thrush singing powerfully in the background, or even a garden warbler or chiffchaff. That sort of advice would have been a real help to me. It would have painted a picture in my mind of what a summer afternoon in my garden might look and sound like.

Here are some of the basic questions I wish someone had answered for me when I first set out to rediscover the birds around me, and I was probably a bit too embarrassed to ask:

When do birds sing? Birds only tend to sing in the spring and early summertime. They stop singing in July and August at the start of the moult, which is when they shed their old feathers and grow new ones for the winter months. In winter the birds are largely silent (apart from robins), though you will probably hear their sharp alarm calls at this time of year, as they warn other birds of nearby predators.

Why do birds sing? Male birds sing because they want to establish their territories and attract a mate. Female birds don't tend to sing, except for a few species like wrens and robins. Some naturalists believe that birds also sing for the sheer damn pleasure of it.

How long do birds live? The average lifespan of most songbirds in the wild is between two and five years. They are usually eaten, freeze or starve to death. In captivity, though, birds will live for much longer. One robin was recorded as living for eleven years in captivity, and the oldest blue tit lived for an astonishing twenty-one years.

What birds can I expect to see at different times of year? Here is a run-through of the seasons that will show you what to look and listen out for at different times of year and where. It is by no means an exhaustive list. It is based on my own experiences in rural and urban Britain.

SPRING

The first birds to start singing in spring are the chaffinches and blackbirds, some as early as February. It will be easy to spot them because the leaf will not yet be on the tree. Song thrushes and mistle thrushes will have started to sing in December and January.

Spring migration: From March onwards a steady flow of small warblers will begin to arrive from Africa – the nightingale is the most famous one, which sticks largely to south-east England. Other arriving warblers include: blackcap warblers, garden warblers, reed and sedge warblers and willow warblers, all of which are widely distributed across Britain and Europe. Their songs will bring the hedgerows to life at this time of year – in particular, the chiffchaff, with its persistent onomatopoeic call.

The cuckoo arrives from Africa in April and will sing its springtime 'cuckoo-cuckoo' refrain till the end of June when it returns to Africa. Cuckoos use other birds' nests – like dunnocks or meadow pipits – to rear their young. The young cuckoo will push out the other bird's eggs and young after twelve days.

These are by no means the only birds that migrate to Britain to breed here in the spring. Others include: whitethroats, flycatchers, yellow wagtails and turtle doves.

The dawn chorus: One of the great miracles of nature. This is a time of year when male birds compete for territories and look for a mate. They announce their presence via shimmering feathers and loud sweet song. The dawn chorus will start with quite muted tones from about March and reach a deafening crescendo by about May. It tails off a bit in June and disappears altogether in July when the breeding season is over and the moult begins.

Breeding: Birds will start to scope out suitable nesting sites from about December and start building nests at around the end of February. From March and up until the end of June all species will be singing, nest building and breeding. Most birds will rear more than one brood and some robins might even rear four clutches of chicks in a season. In general, for the small birds it takes about two weeks for eggs to hatch and another two weeks before the chicks can then fledge. It's slightly longer for larger birds. Chief predators of birds' nests at this time are magpies, crows and jays, which eat the eggs and newborn young. Domestic cats also kill tens of millions of songbirds each year.

SUMMER

This is the time of year when you will see swallows, house martins and swifts in abundance. They begin to arrive from their African wintering grounds in April and May but by June they should be settled and breeding. A great place to watch them is near to water, when they skim the surface hunting for flying insects. Swallows nest in sheds and barns, house martins build conical muddy nests under the eaves of buildings, and swifts nest high up in old buildings like church towers.

Some birds of prey also migrate. If you see lots of swifts in the sky, you might also see a hobby, which is a medium-sized bird of prey with a grey back and mottled white and black breast. It eats swifts, and in flight it looks remarkably similar to its prey. Hobbies mainly stick to the south of Britain.

If you are in the open country in spring and early summer, listen out for the skylark always. They hover thirty to fifty feet above the ground and sing loudly and persistently. You might also hear the distinctive call of the yellowhammer, which sounds a bit like someone saying 'a-little-bit-of-bread-with-no-cheeeeeese'. Meadow pipits will also spring up from the ground everywhere making a *pcheep*, *pcheep*, *pcheep* sound. The mournful sharp mew of a circling buzzard will also be likely to ring out.

If you are in a park or garden at this time of year, the most common songs you will hear will be a blackbird's hearty notes, a wren's ecstatic trill, a greenfinch's throaty *churrrr*, a chaffinch singing a cheerful – if a little grating – tune, a great tit energetically repeating 'teach-er, teach-er, teach-er' over and again, and the more subtle songs of the blue tits and robins. If you are lucky, you might also hear a song thrush's loud fluting melody, which repeats a note over and again, and very often a great spotted woodpecker will be 'drumming' nearby as a means to attract a mate. A constant soundtrack against the background of all these other birds will be the cheerful chirruping of house sparrows, morning noon and nearly night, season over season, year after year.

It will be hard to tell all these songs and sounds apart at first, but it gets easier over time and there are plenty of apps and websites that will help.

AUTUMN

In autumn migration happens in the opposite direction. The summer migrants will have returned to Africa, but large swathes of birds will migrate from the 'frozen north' of the world (places like Russia, Iceland and Scandinavia) down to northern and southern Europe to seek shelter, food and milder weather.

One particular species to look out for are fieldfares and redwings. Hundreds of thousands of these small thrushes migrate each year and will be found hunting for food in arable fields and hedgerows all over Britain. The fieldfare is a slightly meatier version of a song thrush with a blue-grey head and chestnut back. Redwings are slightly smaller but with little red-orange flashes under their wings.

As well as unfamiliar species, like the fieldfares, Britain's native species will also be joined by many thousands of cousins from the north – like chaffinches, skylarks and starlings. In addition many varieties of duck will arrive in autumn to join the common mallards on our ponds and lakes – like wigeon and teal. Wading birds, too, like woodcock, snipe and curlew, will all move south for the winter.

The best place to spot these large flocks of migratory birds is on the east coast of Britain from about October onwards – in particular the brent geese that arrive in huge numbers all the way from Siberia. The November full moon is called the woodcock moon because that is when most of the wintering woodcock cross the North Sea and arrive on the eastern shores of Britain. The woodcock is a dumpy-looking creature with short legs and a long, probing beak. It is the best-camouflaged of all the waders; in its natural winter woodland habitat it looks just like a pile of rotting leaves. But it's easily spotted in flight because it dances, flops and zigzags through the air.

Autumn is also the time of year when tawny owls hoot the loudest. One calls out 'tu-whit' and the other replies 'tu-whu'.

WINTER

In the park or garden: The best time to put out bird feeders is in winter, when birds need food the most. It is not necessary to put food out for the birds in spring when there is plenty for them to eat in the trees and hedgerows. In fact, putting out food in late spring and summer can do more harm than good – bird feeders disturb natural nesting patterns and attract predators, and can spread disease if not cleaned regularly.

Avoid buying an actual bird table because these can too easily turn into convenient buffet bars for any local domestic cat or bird of prey on the prowl, in particular, sparrowhawks – a compact and devastatingly effective small hawk that hunts in confined spaces, like gardens, and eats over 100 different species, with a notable penchant for small birds.

To mitigate the impact of these hawks and cats I find it is best to hang nut-filled wire feeders high up in trees – with plenty of space all around so birds can see predators coming. And also not far from a thick bush or ivy-clad tree they can shelter in when not at the feeder. It's also sensible to buy feeders that are squirrel-proof, which can keep out larger birds like jackdaws, crows and magpies.

Here is a list of the most abundant birds you will attract to your feeder (and what they look like):

- **Robin** – classic red-breasted friend.
- **House sparrow** – squadrons of them in a riot of browns, beiges, greys and blacks.
- **Blackbird** – proprietorial thrush with a deep black sheen and yellow beak*.
- **Blue tit** – delicate bundles with a yellow breast, blue back and a blue cap.
- **Great tit** – muscular bundles with green back, yellow breast and black cap.

* Females are a sturdy brown in colour.

- **Coal tit** – rather like someone shrank a great tit by half.
- **Long-tailed tit** – all tail and no tit. Tiny pinky-grey, brown and black body. Also known as a bottle tit or mumruffin.
- **Chaffinches** – shimmering copper breast, black and white wing flashes and a steel-grey cap*.
- **Greenfinches** – rather like someone spray-painted a chaffinch golden green.
- **Goldfinches** – red, white and black head, brown and buff body with gold wing flashes.
- **Siskin** – a little yellow and black finch.
- **Dunnock** – if John Le Mesurier had been a bird. Self-effacing grey and streaky brown creature.
- **Wren** – tiny wee flitting bird with a dappled chestnut body and a cocked tail.
- **Nuthatch** – presents a perfect right angle on the feeder. Long beak and black eye stripe with a blue-grey back and orange breast.
- **Great spotted woodpecker** – dressed in splendid white shirt and black tailcoat, with a red splodge on its head and bum.

Birds of the open country: Birds tend to form huge flocks in winter as a means to find food, protect themselves from predators and keep warm. It can be spectacular to watch. One of the more moving sights to look out for,

---

* Females look like you put the male through a non-colour-fast wash.

from about November, is a murmuration of starlings: these flocks of many thousands of starlings float in the air like giant shoals of fish. Starlings are a thrush-sized bird of town and country, and stand out with their black, green and violet iridescent wings and body.

When walking across farmland you might also see large flocks of goldfinches feeding in thistle-strewn fields, coveys of native grey partridge or great parliaments of rooks foraging for worms and grubs. There may also be groups of many different species mobbing together on arable fields – like finches, sparrows, buntings, pipits, linnets and larks – looking for the gleanings of last year's harvest. Winter is a wonderful time to be in wild and open country.

One of the finest sights of the open country in winter, though, is of the kestrel hunting. Kestrels are easy to tell apart from, say, a buzzard or a red kite, because they are the only bird of prey that hovers. The kestrel – a small, elegant raptor with a mottled chestnut back and black wingtips – will hover perfectly still in the air above its prey – usually a mouse or a vole. Buzzards are much larger than kestrels, with a four-foot wingspan, and undertake long circular soaring flights as they scan the ground for rabbits, hares or mice. Red kites are slightly larger than buzzards and very easily spotted because of their sharply forked tail.

But the biggest, most magnificent, raptor of them all is the eagle. The best way to catch a glimpse of one is by going to the wilds of Wales or Scotland (at any time of year). Britain is home to two species of eagle: the white-tailed eagle (or sea eagle) and the golden eagle. The sheer scale of them, with their eight-foot wingspans, will take your breath away. And if you find yourself asking, when you see it, is that an eagle or a buzzard? It will be a buzzard. Because you *know* when you've seen an eagle.

For those who want to get a keener sense of the wild world around them, then winter is the perfect time to seek solace in, and knowledge of, nature. The landscape is bare, the birds are hungry and it is the best time of year to get close to them.

Acknowledgements

Here are the life-saving people I have to thank for giving flight to my birds. Without Gordon Wise at Curtis Brown there would be no book. He breathed life into a vague and half-formed idea, presented to him by a vague and half-formed author. He gave me a chance. And I will never be able to thank him enough for his incredible foresight, sound advice and infectious enthusiasm (and a bloody good title too). The idea that a book I had written might one day be published by part of the Penguin family was such a distant fantasy that I never once even dared entertain the idea. And yet it has happened. And it's down to Gordon. And thank you, too, to all the team at Curtis Brown who worked so hard to get this project off the ground.

I would like to thank Daisy Meyrick in particular for letting me believe that any of this was possible in the first place – and for being such a good friend throughout the process. Our grandfathers would be proud.

The editorial team at Penguin Michael Joseph comprehensively exceeded all my expectations. Their kind words and gentle tweaking have encouraged and inspired me in equal measure. Thank you, Charlotte Hardman, Beatrix McIntyre and Paula Flanagan, for all your support and wisdom. Thank you, Jennie Roman, for your expert copy-editing, intelligent bird-based observations and joy-inducing enthusiasm. I hope you found that tawny owl roost in the end? But most especially, thank you, Ariel Pakier; I think I must have won life's Editor Lottery, because I have never worked with someone who was better able to inspire my confidence and trust. Through her diligence, hard work and endless patience, Ariel was able to tease out of me stories and feelings from my past I would never have been able to articulate on my own. She gave this book the beating heart that it needed.

Olivia Thomas, Ella Watkins and Sophie Shaw: you launched my book into the real world, and I can't thank you enough for all your hard work and kindness.

I would like to thank my father, Peter Corbett, for being like no other and for being so patiently accepting of the colourful stories I told about him in the book – not all of them showing him up in the most glowing of lights. 'Well, if it helps you sell more copies.' Thank you, Dad. You were the best and most adoring husband Mum could ever have had. And to Katie Brooke and Richard Corbett, too: your encouragement and support is a major reason why I was able to write this book at all. I could not have done it without your backing. You kept reassuring me that Mum would be so proud and that kept me going.

Andrew Pott – Uncle Andrew – it was you who properly introduced me, over many happy lunches, to a family I had never met: my own. The stories and photographs you showed me of your childhood turned a hazy black-and-white, two-dimensional perception of my heritage into glorious 3D.

And Suzie Neate – Saint Suzie – the beautiful picture

you drew for me of a jenny wren – with the lovely quote I have since included in the book – sits on my desk next to a picture of Mum and inspires me every day, as do the beautiful old books on birds that you so kindly gave me. In fact, *British Birds' Eggs and Nests* (price one shilling), written all those years ago, provided me with the most wondrous descriptions and old-fashioned names, and was the inspiration for the opening of the chapter on the house sparrow.

To Jim Corbett, thank you for the limitless optimism, enthusiasm and hard work you put into keeping Lochbuie the place the family all want and remember it to be. It could not have survived without you and Aunt Patience at the helm. And thank you too to cousin Tom and wife Flora for breaking your backs so that the rest of us can swan up from the south whenever we feel like it and reap all the benefits of your incredible hard work.

I would also like to thank my old friend Charlie Tryon – the keenest fisherman I know – who taught me to fish for trout on quiet silver streams all those years ago. Without those fishing trips to the Wiltshire Avon, meeting the legendary Uncle Aylmer and Charlie's seemingly endless knowledge of the wildlife humming, singing and buzzing around us, my love for chalk streams would never have been born and the chapter on the kingfisher would never have been written. And thank you, too, Jamie Watherston, for resurrecting my passion for fly fishing in later life on our many hilarious trips to the Rivers Kennet, Avon and Nadder – and the pretty pubs that sit contentedly on their cowslip banks.

I also want to mention all the staff and benefactors of SongBird Survival, who have worked so tirelessly, and with very little money or thanks, to try to get to the bottom of the worrying decline in so many of our once abundant bird species. Please keep up your amazing work.

Arthur and Teddy, my darling boys, in my lower moments, when I just wanted to chuck in the towel and give up trying, I thought of you two and carried on. And, Mary Corbett, what can I say? You make all this possible and give me the confidence to carry on, no matter how dark the day.

And finally I would like to thank the birds because I could not cope without you.

Permissions Credits